新一代 信息技术
"十三五"系列规划教材

MySQL
数据库应用
实战教程 慕课版

◆ 刘刚 苑超影 编著

人民邮电出版社
北京

图书在版编目（CIP）数据

MySQL数据库应用实战教程：慕课版 / 刘刚，苑超影编著. -- 北京：人民邮电出版社，2019.7
新一代信息技术"十三五"系列规划教材
ISBN 978-7-115-48900-5

Ⅰ. ①M… Ⅱ. ①刘… ②苑… Ⅲ. ①SQL语言—程序设计—高等学校—教材 Ⅳ. ①TP311.132.3

中国版本图书馆CIP数据核字(2018)第158858号

内 容 提 要

　　MySQL是目前流行的采用客户端/服务器的关系型SQL数据库管理系统之一。本书利用大量案例深入浅出地介绍了使用MySQL管理数据库的基础知识。本书分为9章，分别介绍了MySQL概述、MySQL数据类型、MySQL常用操作、MySQL查询、MySQL函数和存储过程、MySQL高级特性、索引、综合案例——图书管理系统、MySQL管理。本书实战案例丰富，内容全面，配套慕课课程，由一线程序员手把手教学，支持移动和在线学习。

　　本书适合作为高等院校、高职高专MySQL程序设计相关课程的教材，也可供MySQL学习者阅读参考。

◆ 编　著　刘　刚　苑超影
　责任编辑　桑　珊
　责任印制　马振武

◆ 人民邮电出版社出版发行　北京市丰台区成寿寺路11号
邮编 100164　电子邮件 315@ptpress.com.cn
网址 https://www.ptpress.com.cn
北京盛通印刷股份有限公司印刷

◆ 开本：787×1092　1/16
印张：18.25
字数：547千字
2019年7月第1版
2025年2月北京第10次印刷

定价：54.00元

读者服务热线：(010)81055256　印装质量热线：(010)81055316
反盗版热线：(010)81055315

前言
Foreword

本书全面贯彻党的二十大精神，以社会主义核心价值观为引领，传承中华优秀传统文化，坚定文化自信，使内容更好体现时代性、把握规律性、富于创造性。

MySQL 数据库是当前最主流的关系型数据库之一，由瑞典MySQL AB公司开发，目前属于Oracle公司。它免费、开源、体积小、速度快，功能齐全，使用便捷，可运行于Windows和Linux操作系统中，搭配PHP和Apache可组成良好的开发环境。因此，一般中小型网站的开发常选择MySQL作为网站数据库。

本书通过92个实战演练，详细讲解了MySQL数据库的知识和使用方法，配套慕课资源，由一线程序员小刚老师详细讲解，手把手教学。慕课资源可扫描书中二维码移动观看，也可登录人邮学院进行系统化学习。

如何学习慕课版课程

本课程依托人民邮电出版社自主开发的在线教育慕课平台——人邮学院（www.rymooc.com），该平台为学习者提供优质、海量的课程，课程结构严谨，用户可以根据自身的学习程度，自主安排学习进度。人邮学院为每一位学习者提供完善的一站式学习服务（见图1）。

图1　人邮学院首页

为了使读者更好地完成慕课的学习，现将本课程的使用方法介绍如下。

1. 用户购买本书后，找到粘贴在书封底上的刮刮卡，刮开，获得激活码（见图2）。
2. 登录人邮学院网站（www.rymooc.com），或扫描封面上的二维码，使用手机号码完成网站注册（见图3）。

图2　激活码

图3　注册人邮学院网站

3. 注册完成后，返回网站首页，单击页面右上角的"学习卡"选项（见图4），进入"学习卡"页面（见图5），输入激活码，即可获得该慕课课程的学习权限。

图4　单击"学习卡"选项　　　　　　图5　在"学习卡"页面输入激活码

4. 用户可随时随地使用计算机、平板电脑、手机学习本课程的任意章节，根据自身情况自主安排学习进度（见图6）。

5. 在学习慕课课程的同时，阅读本书中相关章节的内容，巩固所学知识。本书既可与慕课课程配合使用，也可单独使用，书中主要章节均放置了二维码，用户扫描二维码即可在手机上观看相应章节的视频讲解。

6. 书中配套的PPT、源代码等教学资源，用户也可在该课程的首页找到相应的下载链接。

关于使用人邮学院平台的任何疑问，都可登录人邮学院咨询在线客服，或致电：010-81055236。

图6　课时列表

小刚老师简介

- 一线项目研发、设计、管理工程师，高级项目管理师、项目监理师，负责纪检监察廉政监督监管平台、国家邮政局项目、政务大数据等多个国家级项目的设计与开发
- 极客学院、北风网金牌讲师
- 畅销书《微信小程序开发图解案例教程（附精讲视频）》《小程序实战视频课：微信小程序开发全案精讲》《Axure RP8原型设计图解微课视频教程（Web+App）》作者

全部案例源代码、素材、最终文件、电子教案可登录人邮教育社区（www.ryjiaoyu.com.cn）下载使用。

编著者

2022年12月

目录
Contents

第1章　MySQL概述　1
1.1　MySQL的发展历史　2
　　精讲视频——MySQL的发展历史　2
1.2　了解关系数据库　3
　　精讲视频——了解关系数据库　3
　　1.2.1　什么是关系数据库　3
　　1.2.2　关系模型　4
　　1.2.3　常用的关系型数据库　4
　　1.2.4　SQL结构化查询语言　4
1.3　MySQL的工作原理　5
　　精讲视频——MySQL简介　5
1.4　MySQL数据库存储引擎　6
　　精讲视频——MySQL数据库存储引擎　6
　　1.4.1　InnoDB引擎　6
　　1.4.2　MyISAM引擎　6
　　1.4.3　MEMORY引擎　7
　　1.4.4　CSV引擎　7
　　1.4.5　ARCHIVE引擎　7
　　1.4.6　BLACKHOLE引擎　8
　　1.4.7　存储引擎特点对比　8
1.5　实战：安装MySQL数据库　9
　　精讲视频——实战：安装MySQL数据库　9
　　1.5.1　在Windows 环境下
　　　　　安装MySQL　9
　　1.5.2　Linux环境下安装　13
1.6　小结　14

第2章　MySQL数据类型　15
　　精讲视频——MySQL数据类型　16
2.1　数值类型　16
　　精讲视频——数值类型　16
2.2　字符串类型　17
　　精讲视频——字符串类型　17
2.3　日期时间类型　18

　　精讲视频——日期时间类型　18
2.4　复合类型　18
　　精讲视频——复合类型　18
2.5　如何选取数据类型　18
　　精讲视频——如何选取数据类型　18
2.6　小结　19

第3章　MySQL 常用操作　20
3.1　数据库用户管理　21
　　精讲视频——数据库用户管理　21
　　3.1.1　连接MySQL　21
　　　　实战演练——数据库本地连接　21
　　　　实战演练——数据库远程连接　21
　　3.1.2　新增用户　22
　　　　实战演练——新增用户　22
　　3.1.3　修改用户密码　23
　　　　实战演练——修改用户密码　23
3.2　数据库操作　24
　　精讲视频——数据库操作　24
　　3.2.1　查看数据库　24
　　　　实战演练——查看数据库　24
　　3.2.2　创建数据库　25
　　　　实战演练——创建数据库　25
　　3.2.3　使用数据库　26
　　　　实战演练——使用数据库　26
　　3.2.4　删除数据库　26
　　　　实战演练——删除数据库　26
3.3　表操作　27
　　精讲视频——表操作　27
　　3.3.1　创建表　27
　　　　实战演练——创建表　27
　　3.3.2　查看表结构　28
　　　　实战演练——查看表结构　28
　　3.3.3　复制表　29

🔗 实战演练——复制表结构、数据、主键、索引	29
🔗 实战演练——复制表结构、数据	32
3.3.4 临时表和内存表	34
🔗 实战演练——临时表和内存表	34
3.4 数据操作	35
🔗 精讲视频——数据操作	35
3.4.1 插入（INSERT）、查询（SELECT）	35
🔗 实战演练——数据插入和查询	36
3.4.2 修改记录（UPDATE）	37
🔗 实战演练——修改记录	38
3.4.3 删除记录（DELETE）	39
🔗 实战演练——删除记录	39
3.4.4 对查询结果排序（ORDEY BY）	40
🔗 实战演练——查询结果排序	41
3.4.5 对查询结果分组（GROUP BY）	42
🔗 实战演练——查询结果分组	43
3.4.6 设置分组条件（HAVING）	44
🔗 实战演练——设置分组条件	45
3.4.7 限制查询数量（LIMIT）	46
🔗 实战演练——数据检索	46
3.5 字段操作	48
🔗 精讲视频——字段操作	48
3.5.1 设置为主键	48
🔗 实战演练——主键设置	48
3.5.2 设置为复合主键	50
🔗 实战演练——复合主键设置	50
3.5.3 添加字段	51
🔗 实战演练——添加字段	51
3.5.4 改变字段类型	52
🔗 实战演练——改变字段类型	52
3.5.5 字段重命名	53
🔗 实战演练——字段重命名	53
3.5.6 字段设置默认值	54
🔗 实战演练——字段设置默认值	54
3.5.7 设置自增字段	56
🔗 实战演练——设置自增字段	56

3.6 客户端操作数据库	58
🔗 精讲视频——客户端操作数据库	58
🔗 实战演练——使用客户端	58
3.7 小结	61
第4章 MySQL查询	**62**
4.1 基本查询语法	63
🔗 精讲视频——基本查询语法	63
🔗 实战演练——查询数据	63
4.2 数据过滤	67
🔗 精讲视频——数据过滤	67
4.2.1 基本查询过滤	67
🔗 实战演练——基本查询过滤	67
4.2.2 条件查询过滤	68
🔗 实战演练——条件查询过滤	68
4.2.3 模糊查询过滤	72
🔗 实战演练——模糊查询过滤	72
4.2.4 字段控制查询过滤	73
🔗 实战演练——字段控制查询过滤	73
4.2.5 正则表达式查询过滤	76
🔗 实战演练——使用正则表达式	76
4.3 子查询	80
🔗 精讲视频——子查询	80
4.3.1 什么是子查询	80
4.3.2 按返回结果分类的子查询	80
🔗 实战演练——表子查询	80
🔗 实战演练——行子查询	81
🔗 实战演练——列子查询	82
🔗 实战演练——标量子查询	84
4.3.3 按对返回结果的调用方法分类的子查询	85
🔗 实战演练——WHERE型子查询	85
🔗 实战演练——FROM型子查询	86
🔗 实战演练——exists型子查询	87
4.4 聚合函数	90
🔗 精讲视频——聚合函数	90
4.4.1 AVG()函数	90
🔗 实战演练——AVG()函数	90
4.4.2 COUNT()函数	91
🔗 实战演练——COUNT()函数	91
4.4.3 MAX()/MIN() 函数	92

	实战演练——MAX()/MIN()函数	92
	4.4.4 SUM()函数	93
	实战演练——SUM()函数	94
4.5	高级查询	95
	精讲视频——高级查询	95
	4.5.1 内连接查询	95
	实战演练——内连接查询	95
	4.5.2 外连接查询	99
	实战演练——外连接查询	99
	4.5.3 自然连接查询	101
	实战演练——自然连接查询	102
	4.5.4 交叉连接查询	104
	实战演练——交叉连接查询	104
	4.5.5 联合查询	107
	实战演练——联合查询	107
4.6	小结	110

第5章 MySQL函数和存储过程 111

5.1	MySQL流程控制函数	112
	精讲视频——MySQL流程控制函数	112
	实战演练——使用流程控制函数	112
5.2	MySQL常用函数	114
	精讲视频——MySQL常用函数	114
	5.2.1 数学函数	114
	实战演练——使用数学函数	115
	5.2.2 字符串函数	117
	实战演练——使用字符串函数	118
	5.2.3 日期和时间函数	120
	实战演练——使用日期和时间函数	122
	5.2.4 系统信息函数	125
	实战演练——使用系统信息函数	125
	5.2.5 加密函数	127
	5.2.6 格式化函数	128
5.3	自定义函数	131
	精讲视频——自定义函数	131
	5.3.1 函数的基本语法	131
	5.3.2 创建不带参数的自定义函数	133
	实战演练——创建不带参数的自定义函数	133
	5.3.3 创建带参数的自定义函数	134

	实战演练——创建带参数的自定义函数	134
5.4	存储过程	135
	精讲视频——存储过程	135
	5.4.1 存储过程的基本语法	135
	5.4.2 创建不带参数的存储过程	137
	实战演练——创建不带参数的存储过程	138
	5.4.3 创建带有IN类型参数的存储过程	139
	实战演练——创建带有IN类型参数的存储过程	139
	5.4.4 创建带有IN和OUT类型参数的存储过程	142
	实战演练——创建带有IN和OUT类型参数的存储过程	142
	5.4.5 创建带有多个OUT类型参数的存储过程	144
	实战演练——创建带有多个OUT类型参数的存储过程	144
	5.4.6 创建带有INOUT类型参数的存储过程	145
	实战演练——创建带有INOUT类型参数的存储过程	145
	5.4.7 创建IF语句的存储过程	147
	实战演练——创建IF语句的存储过程	147
	5.4.8 创建CASE语句的存储过程	149
	实战演练——创建CASE语句的存储过程	149
	5.4.9 创建while循环语句的存储过程	151
	实战演练——创建while循环语句的存储过程	151
5.5	自定义函数和存储过程的区别	152
	精讲视频——自定义函数和存储过程的区别	152
	5.5.1 自定义函数和存储过程的区别	152
	5.5.2 存储过程的使用建议	152
5.6	小结	153

第6章 MySQL 高级特性 154

| 6.1 | 视图 | 155 |

	精讲视频——视图	155
6.1.1	什么是视图	155
6.1.2	创建视图	155
	实战演练——创建视图	155
6.1.3	修改视图	160
	实战演练——修改视图	160
6.1.4	更新视图数据	162
	实战演练——更新视图数据	162
6.1.5	删除视图和数据	166
	实战演练——删除视图和数据	166
6.2	游标	168
	精讲视频——游标	168
6.2.1	游标的使用	168
	实战演练——使用游标	169
6.2.2	游标的WHILE循环	171
	实战演练——WHILE循环的游标	171
6.2.3	游标的REPEAT循环	175
	实战演练——REPEAT循环的游标	175
6.2.4	游标的LOOP循环	177
	实战演练——LOOP循环的游标	177
6.3	触发器	179
	精讲视频——触发器	179
6.3.1	创建触发器	179
	实战演练——创建触发器	180
6.3.2	NEW和OLD关键字	182
	实战演练——使用NEW和OLD关键字	182
6.3.3	查看和删除触发器	185
6.3.4	INSERT型触发器	185
	实战演练——INSERT型触发器	185
6.3.5	UPDATE型触发器	188
	实战演练——UPDATE型触发器	188
6.3.6	DELETE型触发器	190
	实战演练——DELETE型触发器	190
6.4	小结	192

第7章 索引 193

7.1	索引的基本语法	194
	精讲视频——索引的基本语法	194
7.1.1	创建索引	194
7.1.2	查看索引	195
7.1.3	删除索引	195

	实战演练——给用户表user添加和删除索引	196
7.2	常见的查询算法	198
	精讲视频——常见的查询算法	198
7.3	索引的数据结构	199
	精讲视频——索引的数据结构	199
7.3.1	B-Tree数据结构	199
7.3.2	B+Tree数据结构	199
7.4	索引实现原理	200
	精讲视频——索引实现原理	200
7.4.1	MyISAM的索引实现	200
7.4.2	InnoDB的索引实现	200
7.4.3	MEMORY的索引实现	201
7.5	索引的应用	201
	精讲视频——索引的应用	201
7.5.1	创建表及添加索引	201
	实战演练——创建学生表student并添加索引	202
7.5.2	explain分析索引	204
	实战演练——用explain分析student表索引	204
7.5.3	索引使用策略	219
7.5.4	索引会用了吗	223
	实战演练——联合索引应用	224
7.6	索引的类型	233
	精讲视频——索引的类型	233
7.6.1	主键索引	233
	实战演练——主键索引	233
7.6.2	普通索引	234
	实战演练——普通索引	234
7.6.3	唯一索引	235
	实战演练——唯一索引	235
7.6.4	单列索引和联合索引	236
	实战演练——单列索引和联合索引	236
7.6.5	聚簇索引和非聚簇索引	237
7.6.6	覆盖索引	237
7.6.7	重复索引和冗余索引	238
7.7	索引不能使用	239
	精讲视频——索引不能使用	239
7.7.1	不等于（<>、!=）	239

	实战演练——不等于不能使用索引	239
	8.6.1 借书预约管理	257
7.7.2	前导模糊查询（%xx%）	241
	8.6.2 借书登记管理	258
	实战演练——前导模糊查询不能使用索引	241
	8.6.3 还书管理	258
	8.6.4 图书遗失登记管理	258
7.7.3	比较不匹配的数据类型	242
8.7	视图管理	259
	实战演练——比较不匹配的数据类型不能使用索引	243
	精讲视频——视图管理	259
	8.7.1 用户信息查询视图	259
7.7.4	OR连接条件	244
	8.7.2 用户借阅图书查询视图	259
	实战演练——OR连接条件	244
	8.7.3 用户还书查询视图	260
7.7.5	条件表达式前使用函数	246
8.8	小结	260
	实战演练——条件表达式前使用函数	246
第9章	**MySQL管理**	**261**
7.8	索引的利弊及建立原则	248
9.1	用户管理	262
	精讲视频——索引的利弊及建立原则	248
	精讲视频——用户管理	262
7.9	小结	248
	9.1.1 创建用户	262
	9.1.2 修改用户	263
第8章	**综合案例——图书管理系统**	**249**
	9.1.3 删除用户	266
8.1	需求管理	250
9.2	权限管理	268
	精讲视频——需求管理	250
	精讲视频——权限管理	268
8.2	数据库设计	250
	9.2.1 权限授权语法	268
	精讲视频——数据库设计	250
	实战演练	270
8.3	创建数据库	252
	9.2.2 账户权限体系	272
	精讲视频——创建数据库	252
	9.2.3 权限授予原则	273
	8.3.1 建表语句	253
9.3	表空间管理（InnoDB）	275
	8.3.2 初始化数据	255
	精讲视频——表空间管理（InnoDB）	275
8.4	用户信息管理	256
9.4	备份与还原	277
	精讲视频——用户信息管理	256
	精讲视频——备份与还原	277
	8.4.1 用户管理	256
	9.4.1 备份数据	277
	8.4.2 部门管理	256
	9.4.2 还原数据	278
8.5	图书管理	256
9.5	主从同步配置	279
	精讲视频——图书管理	256
	精讲视频——主从同步配置	279
	8.5.1 新增图书分类	256
	9.5.1 主数据库配置	279
	8.5.2 新增图书	257
	9.5.2 从数据库配置	280
8.6	借书管理	257
9.6	小结	282
	精讲视频——借书管理	257

第1章
MySQL概述

本章要点

MySQL的发展历史
了解关系数据库
MySQL的工作原理
MySQL数据库存储引擎
实战：安装MySQL数据库
小结

■ MySQL是采用客户端/服务器的关系型SQL数据库管理系统，它具有跨平台性和可移植性，可以轻松、简单地运行在多种操作系统上，如Windows、Linux操作系统等。本章介绍MySQL的基础知识，并带大家安装MySQL数据库。

MySQL数据库是开放源码的，允许有兴趣的爱好者去查看和维护源码，大公司或者有能力的公司还可以继续对其进行优化，做成适合自己公司的数据库。最重要的一点是，相较于Oracle数据库的商用收费，MySQL允许各大公司免费使用，并且在被甲骨文公司收购后，不断地进行优化，性能提升接近30%，已成为小公司或者创业型公司首选的数据库，市场占有率也逐渐扩大，如图1.1所示。

图1.1　2017年11月数据库占比

1.1　MySQL的发展历史

精讲视频

MySQL的发展历史

MySQL最初是由一个名叫蒙蒂·维德纽斯（Monty Widenius）的小职员为一个叫TcX的公司设计的一款底层面向报表的存储引擎工具——Unireg。

在1985年，Monty和几个志同道合的朋友在瑞典成立了一家公司，也就是MySQL AB的前身。该公司最初也不是致力于做数据库产品的，只是因为工作过程中需要一个数据库，可是又没有合适的数据库选择，就决定自己开发一个数据库。为了满足瑞典的一些大型零售商不断增长的数据服务需求，并为复杂的系统提供数据仓库服务，他们自己设计了一个索引顺序存取数据算法，开发了高查询性能的数据引擎，也就是ISAM数据存储引擎。

1990年，有些用户要求提供SQL支持，于是Monty想将mSQL（mini SQL，即单用户数据库管理系统）代码集成到ISAM存储引擎中，但效果并不好。于是，Monty毅然决定自己重写一个SQL支持。

1996年，MySQL 1.0正式发布，提供的功能非常简单，只有表数据的INSERT（插入）、UPDATE（更新）、DELETE（删除）和SELECT（查询）操作。不过，它采用的许可策略却与众不同，允许免费商用，前提是不能捆绑MySQL一起发布，这为它的后续发展打下了良好的基础。随着MySQL 3.11.1版本的发布，MySQL不仅提供基本的SQL支持，还提供了复杂的查询优化器，尽管如此，MySQL依旧不支持事务、视图、存储过程等特性。

1999~2000年，Monty团队成立了MySQL AB公司，与Sleepcat合作开发了Berkeley DB引擎，MySQL从此支持事务处理。

2000年，MySQL公布了源码，并采用了通用公共许可协议（GNU General Public License, GPL）；4月，MySQL对旧的存储引擎进行了整理，命名为MyISAM，同时支持全文搜索。

2001年，Heikki Tuuri提议集成自己的InnoDB引擎到MySQL中，这个存储引擎不仅支持行级锁，而且还支持事务处理，MySQL和InnoDB正式结合版本为4.0。

2004年，MySQL发了4.1版本，新增了子查询。

2005年10月，MySQL发布了5.0版本，新版本加入了存储过程、触发器、视图等，MySQL逐渐向高性能数据库方向发展。

2008年1月，MySQL被Sun公司收购。

2009年4月，Oracle公司收购了Sun公司，MySQL转入Oracle旗下。

2010年，MySQL 5.5发布，新特性包括半同步的复制以及对SIGNAL/RESIGNAL异常处理功能的支持，同时InnoDB引擎变为MySQL的默认存储引擎，还加强了在企业应用方面的特性。令人庆幸的是，Oracle承诺MySQL 5.5和未来的版本仍然是采用GPL协议的开源数据库。

在 MySQL 5.5 发布两年后，Oracle 宣布 MySQL 5.6 正式版发布，首个正式版版本号为 5.6.10。在 MySQL 5.5 中使用的是 InnoDB 作为默认的存储引擎，而MySQL 5.6则对InnoDB 引擎进行了改造，提供全文索引能力，使 InnoDB 适合各种应用场景。

2015 年，MySQL 5.7 GA 重磅发布，新特性包括160 万只读 QPS，动态修改 Buffer Pool，提高运维效率，优化器 Server 层改进、InnoDB 层优化等。

自MySQL被收购后，Monty Widenius觉得依靠Sun/Oracle来发展MySQL存在问题，于是决定另起分支，于2009年成立了MariaDB项目。

1.2 了解关系数据库

MySQL作为最流行的关系型数据库管理系统之一，具备了容易理解、使用方便、易于维护等特性。那么，什么是关系数据库管理系统？什么是关系模型？关系型数据库有哪些优点呢？

1.2.1 什么是关系数据库

数据库（Database）是以特定数据结构组织，在计算机设备上存储和管理数据的"仓库"。在日常生活和工作中，有许多数据集，需要将它们归档到这样一个仓库中，以便进行数据统计和查询等相关管理。数据库有很多类型，通常根据不同数据组织类型分为层次式数据库、网络式数据库、关系数据库和面向对象的数据库4种，下面主要讲解关系数据库。

层次结构数据库是通过一种有根节点的定向有序树结构（类似于一个倒挂的树）模型建立的数据库，如IMS（Information Management System）；网络式数据库是按照网状数据结构建立的数据库管理系统，记录中允许多层次记录关系；面向对象的数据库是一种新的程序设计方法，将世界中的一切事物都视为对象，整个世界都由相互关联的对象组织而成，如果感兴趣可以查询相关资料。

关系数据库管理系统（Relational Database Management System，RDBMS），是指通过关系模型来组织数据的数据库。关系数据库把世界看成由实体和联系组成。关系模型最初在1970年由IBM的研究员E.F.Codd博士提出，在之后的几十年中，关系模型的概念得到了很好的发展，并且逐渐成为主流数据库架构模型。可以简单地把关系数据库理解为由二维表格建立的数据组织（类似Excel，由行和列组织数据）和二维表格之间的联系构成的数据关联。下面详细解释下关系模型。

1.2.2 关系模型

关系模型是用二维表的形式表示实体以及实体之间联系的数据模型。数据都是以表格的形式存在的，每行对应一个实体的记录，每列对应实体的某种属性，若干行和列构成了整个表数据。

所谓实体就是现实世界中客观存在的，可以是有形的、无形的、具体的或者抽象的事物。

说到实体，不得不说实体关系模型。实体关系模型是能直观表示实体、属性以及和实体间联系的模型，可以通过实体关系图（Entity Relationship Diagram，E-R图）来表示，是用来理解现实生活中的实体关系、建立概念模型非常有效的工具。

例如，某个社交网站和用户之间的关系如图1.2所示。

在图1.2的E-R图中，社交网站和用户代表实体，用矩形表示，社交网站拥有属性：企业性质、网站名、上市，用椭圆形表示；用户拥有属性：姓名、手机号、生日，也用椭圆表示；实体社交网站和实体用户之间的联系，用菱形表示。在实体关系模型中，联系（也称关系）有3种类型：一对一关系（1:1），如用户有一个会员编号，一个会员编号能确定唯一一个用户，他们之间是一一对应的；一对多关系（1:N），如用户可以注册多个登录账号，而注册账号只能被当前的用户使用；多对多关系（N:N），如社交网站可以有多个用户，而用户同样可以拥有多个社交网站。

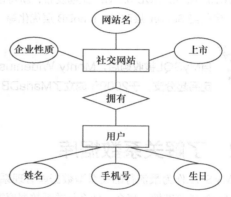

图1.2 社交网站用户关系图

1.2.3 常用的关系型数据库

比较常用的关系型数据库有Oracle、MySQL、SQL Server、IBM DB2、Sybase、Access等，Oracle是收费商用的数据库，提供很好的维护与支持，适用于业务逻辑较复杂、数据量大的大中型项目；MySQL数据库由于体积小、速度快、总体拥有成本低、开放源码，受到很多中小型公司的青睐；SQL Server数据库的功能比较全面、效率高，适用于中型企业或单位的数据库平台。

1.2.4 SQL结构化查询语言

SQL（Structured Query Language）意为结构化查询语言，是一种用于数据库查询和程序设计的语言，可用于操作数据库的查询、存取更新数据以及管理数据库，是通用的数据库查询语言，它具有极大的灵活性和很多强大的功能。虽然SQL是国际标准的关系型数据库管理语言，但事实上，很多流行的数据库都对SQL规范做了些修改扩充，所以，不同的数据库之间不能完全地互通，如MySQL没有top命令，SQL Server没有limit命令等。

1.3 MySQL的工作原理

了解了关系数据库的基础知识，下面我们开始讲解MySQL。

为了理解MySQL的工作原理，我们先看一张经典的架构图，如图1.3所示。

MySQL的内部架构由以下几个部分组成。

（1）编程语言交互接口（Connectors）：指的是不同语言与SQL的交互接口，如Java的JDBC、.Netframework的ODBC。

（2）系统管理和控制工具集合（Management Services & Utilities）：提供管理配置服务、备份还原、安全复制等功能。

MySQL简介

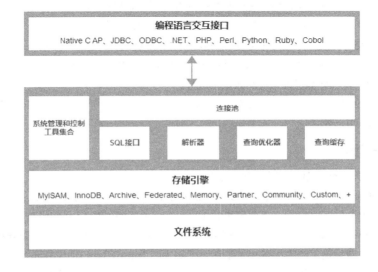

图1.3　MySQL的内部架构

（3）连接池（Connection Pool）：接受客户端的请求，缓存请求，检查内存可利用情况，如果没有可用线程，就创建线程执行任务，有可用线程就重复利用。

（4）解析器（Parser）：解析验证SQL语法，分解SQL成相应的数据结构，以备后面处理。

（5）查询优化器（Optimizer）：对SQL语句进行优化处理，优化执行路径，生成执行树，最终数据库会选择认为最优的方案执行并返回结果。

（6）SQL接口（SQL Interface）：接受用户的SQL命令，并返回结果。

（7）查询缓存（Cache & Buffer）：缓存查询结果。如果SQL查询中命中查询结果，将直接从缓存中返回结果，不再执行SQL分析等操作；没有命中，才会进行后续的解析、查询优化、执行SQL接口，返回结果，同时将结果加入缓存中。

（8）存储引擎（Pluggable Storage Engines）：是MySQL中具体的与文件打交道的子系统，可以看到它是以插件形式存在的，意味着可以自定义存储引擎，这是MySQL很特别的地方。MySQL提供了很多存储引擎，其优势各不一样，有的查询效率高、有的支持事务等，最常用的有MyISAM、InnoDB、BDB。

（9）文件系统（File System）：是存放数据库表数据以及相关配置的地方。

下面举例来说明MySQL的查询过程，如用户要查询具体用户的详情（SELECT * FROM T_USER WHERE ID = 'ID'）。

（1）客户端先发送这条查询命令给MySQL服务器。

（2）服务器会先检查缓存，如果缓存命中，则立即返回缓存中的数据；否则，服务器进行SQL解析、预处理，再通过优化器生成执行计划。

（3）服务器根据生成的执行计划，调用对应引擎的API来执行查询。

（4）将结果返回客户端。

1.4 MySQL数据库存储引擎

MySQL中的数据可以采用不同的技术存储在文件（或内存）中。这些技术都使用不同的存储机制、索引技巧、锁定水平，并且最终提供广泛的、不同的功能和能力。通过选择不同的技术，我们能够获得额外的速度或者功能，从而改善应用的整体功能，每种技术以及配套的相关功能就可以看成是一种数据库存储引擎，MySQL默认配置了许多不同的存储引擎，这些存储引擎可以预先设置或者在MySQL服务器中启用。

精讲视频

MySQL数据库
存储引擎

例如，银行转账交易需要一个支持事务处理的数据库，以确保事务处理不成功时数据的回退能力，这时就不能选用MyISAM存储引擎，因为它是非事务性存储引擎，可以采用InnoDB引擎，这样才能保证银行转账正常进行。

MySQL自身提供的存储引擎有InnoDB（5.5版本后为默认存储引擎）、MyISAM、MEMORY、CSV、ARCHIVE、BLACKHOLE等。

1.4.1 InnoDB引擎

作为默认存储引擎，InnoDB具备以下主要优势。

（1）数据操纵语言（Data Manipulation Language，DML）操作遵循事务的4个特性——原子性（Atomicity）、一致性（Consistency）、隔离性（Isolation）、持久性（Durability），并通过commit、rollback、crash-recovery保障数据的安全。具体来说，crash-recovery就是指如果服务器因为硬件或软件问题而崩溃，不管当时数据是怎样的状态，在重启MySQL后，InnoDB会自动恢复到发生崩溃之前的状态，并回到用户离开的地方。另外，如果数据在磁盘或者内存中损坏，校验机制会提醒当前数据为虚假数据。

（2）具有行级锁和Oracle风格的读一致性，通过一种更改缓存机制对新增、更新和删除进行优化，增加了对用户并发读写操作和性能。

（3）对表进行基于主键的优化查询，每张表都有一个基于主键的聚集索引，以此达到减少磁盘I/O，进而提高了搜索效率和性能。

（4）支持外键约束，检查外键、插入、更新和删除，以确保数据的完整性。

（5）InnoDB引擎提供了专门的缓存池，在内存中缓存了表和索引的数据，常用的数据可以直接从内存中处理，比直接从磁盘获取数据处理速度快。

（6）可以压缩表和相关索引，创建和删除索引以达到提高性能的目的。

（7）快速压缩表空间，并能释放磁盘空间，保证系统能够重用，而不仅仅是腾出空间给InnoDB复用。

（8）创建表时可以通过语句显示指定表引擎：

CREATE TABLE TABLE_NAME (I INT) ENGINE = INNODB;

1.4.2 MyISAM引擎

MyISAM引擎不支持事务、外键，不过访问速度非常快，表的存储分为以下3个文件。

（1）frm文件：存储表定义。

（2）MYD（MYData）文件：存储数据。

（3）MYI（MYIndex）文件：*存储索引*。

它有以下特点。

（1）所有数值类型键值都是以高字节存储的，以便于更好地索引压缩。

（2）每张MyISAM表最多支持$(2^{32})^2$（1.844E+19）行。

（3）每张MyISAM表支持的最大索引数是64，每个索引最多16列。

（4）当表字段是通过自增长（auto_increment）的，索引树节点只会包含一个键，这样可以提高索引的空间利用率。

（5）在INSERT、UPDATE时，MyISAM有内部处理会自动更新auto_increment字段，这使auto_increment处理更快（至少10%）。

（6）当进行混合操作（删除、更新、插入同时进行）时，MyISAM通过自动合并和扩展删除块，减少了行碎片。

（7）MyISAM支持并发插入数据：如果一个表中的数据文件中没有空闲块，则可以在插入数据的同时通过其他线程读取表数据。空闲块是由删除或更新操作时数据长度超过当前行内容长度引起的。

（8）可以通过将数据文件（MYD）和索引文件（MYI）放在不同物理设备的不同目录上，来更快地创建表。

（9）BLOB和TEXT可以被索引，索引列中允许NULL值，不过需要占0～1个字节。

（10）创建表时可以通过语句显示指定表引擎：

CREATE TABLE TABLE_NAME (I INT) ENGINE = MYISAM;

1.4.3 MEMORY引擎

MEMORY引擎又称为HEAP引擎，用来创建特殊用途的表，且内容存储在内存中。将数据存储在内存中，能够实现快速访问和低延迟。

因此，使用MEMORY引擎，在出现数据崩溃、硬件故障等问题时，数据极易丢失；它适合临时态和非关键数据（如会管理或缓存等）的操作。

创建表时可以通过语句显示指定表引擎：

CREATE TABLE TABLE_NAME (I INT) ENGINE = MEMORY;

1.4.4 CSV引擎

当用户创建一个CSV引擎的表时，服务器会在数据库目录中创建一个"表名.frm"的格式文件，同时还会创建一个"表名.csv"的数据文件，该数据文件中的数据是以逗号分隔保存的，它主要用于CSV报表格式的数据存储，应用面比较窄。

缺点：CSV引擎不支持索引，也不支持分区，并且所有列必须指明为NOT NULL。

创建表时可以通过语句显示指定表引擎：

CREATE TABLE TABLE_NAME (I INT) ENGINE = CSV;

1.4.5 ARCHIVE引擎

ARCHIVE引擎用于数据归档，它的压缩比例非常高，适合存储历史数据（前提是不做查询操作），所占的存储空间不到InnoDB引擎的1/10；它支持行级锁实现并发插入操作，却不支持事务，其设计目的在于提供高速插入和压缩功能；另外，它不支持索引。

创建表时可以通过语句显示指定表引擎：

CREATE TABLE TABLE_NAME (I INT) ENGINE = ARCHIVE;

1.4.6 BLACKHOLE引擎

BLACKHOLE引擎是很特别的一种引擎，它的表不存储任何数据，就像是"黑洞"一样。它主要用于充当伪服务器、日志服务器、增量备份服务器等。

创建表时可以通过语句显示指定表引擎：

CREATE TABLE TABLE_NAME (I INT) ENGINE = ARCHIVE;

MySQL还支持其他引擎：MERGE引擎、FEDERATED引擎、EXAMPLE引擎等，可以查看相关文档。

1.4.7 存储引擎特点对比

MySQL在5.5版本之后默认的存储引擎是InnoDB存储引擎，下面对比一下MySQL 5.7版本存储引擎的特点，如表1.1所示。

表1.1 MySQL 5.7版本存储引擎的特点

特点	InnoDB	MyISAM	MEMORY	ARCHIVE
存储限制	64TB	没有	有	没有
事务安全	支持			
锁机制	行锁	表锁	表锁	行锁
B树索引	支持	支持	支持	
哈希索引	支持		支持	
全文索引	支持	支持		
集群索引	支持			
数据缓存	支持		支持	
数据可压缩		支持		支持
空间使用	高	低		非常低
内存使用	高	低	中等	低
批量插入速度	低	高	高	非常高
支持外键	支持			
复制支持	支持	支持	支持	支持
查询缓存	支持	支持	支持	支持
备份恢复	支持	支持	支持	支持

选择存储引擎的建议如下。

（1）MySQL的存储引擎很多，不同的库、不同的表可以选择不同的存储引擎，推荐同一个库用同一种存储引擎，因为不同存储引擎的表之间join操作比较慢。

（2）InnoDB存储引擎提供了具有提交、回滚和崩溃恢复能力的事务安全表，如果需要事务处理、ACID事务支持，则选择InnoDB存储引擎。

（3）Memory将所有数据保存在缓存RAM中，可以提供极快的访问速度。

（4）尽量不要选择MyISAM存储引擎，它只能用单个CPU，内存只能用到4GB，内存里只有索引，而且并发能力差。

1.5 实战：安装MySQL数据库

首先，通过官网下载MySQL，如图1.4所示。

可以选择Windows、Linux、Mac OS等不同的操作系统以及32位或64位系统，和不同版本进行下载，这里使用的是Windows 7系统。本书以Windows MySQL 5.7.20版本为例，后面的所有例子也是基于这个版本。

精讲视频

实战：安装
MySQL数据库

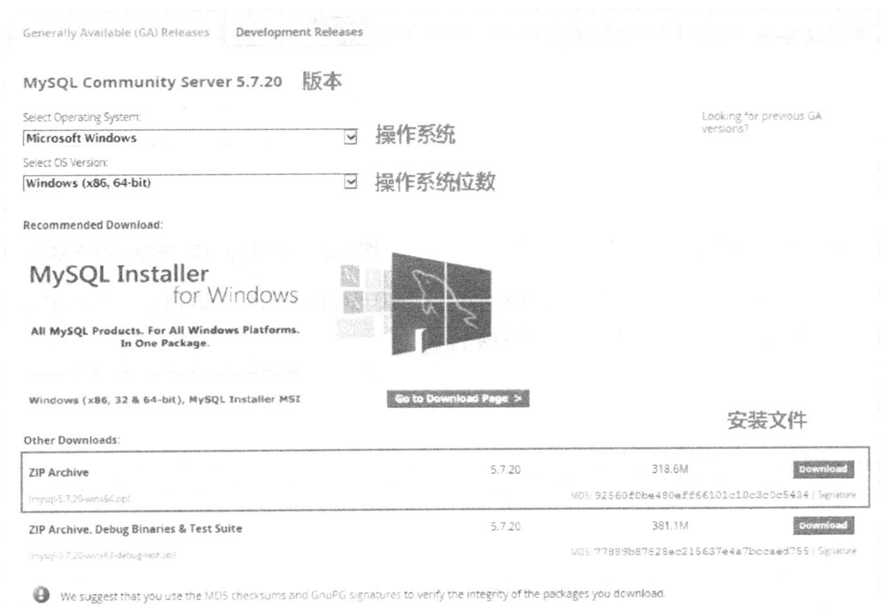

图1.4 MySQL下载页面

1.5.1 在Windows 环境下安装MySQL

（1）从官网下载mysql-5.7.20-winx64 （即Windows 64位系统）版本，将下载的压缩文件放到目标安装目录中，如D:\Program Files\MySQL，开始安装免安装5.7.20版本的MySQL。

（2）如果计算机中有以前的MySQL版本，如果是安装版的，则可以在控制面板中卸载，还需要删除以前MySQL留下来的残留文件；如果是免安装版本，则用mysqld -remove命令删除即可。

（3）解压安装包到当前目标安装目录，解压后的目录如D:\Program Files\MySQL\mysql-5.7.20-winx64。

（4）配置环境变量，选择“我的电脑（计算机）”→“属性”→“高级系统设置”→“环境变量”命令，如图1.5所示，打开“环境变量”窗口。

（5）在弹出的“环境变量”窗口中找到Path，单击“编辑”按钮，添加MySQL文件路径到系统变量Path中，在Path变量值中追加“; D:\Program Files\MySQL\mysql-5.7.20-winx64\bin”，分号不能少，它是用于分割目录的，如图1.6所示。

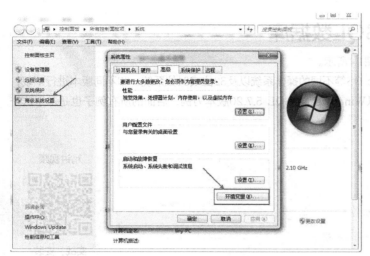

图1.5 打开"环境变量"窗口　　　　　　　　图1.6 添加MySQL路径到系统变量中

（6）添加配置文件，在D:\Program Files\MySQL\mysql-5.7.20-winx64目录下面新建一个记事本文件，修改成my.ini文件格式，打开my.ini文件，添加如下配置。

```
[client]
port=3306
[mysql]
default-character-set=utf8
[mysqld]
port=3306
basedir= D:\Program Files\MySQL\mysql-5.7.20-winx64
datadir= D:\Program Files\MySQL\mysql-5.7.20-winx64\data
character-set-server=utf8
default-storage-engine=innodb
max_connections=512
innodb_flush_log_at_trx_commit=0
innodb_log_file_size         = 1G
innodb_log_files_in_group    = 3
innodb_max_dirty_pages_pct   = 50
innodb_lock_wait_timeout     = 120

sync_binlog=1000

innodb_log_buffer_size=16M
innodb_write_io_threads = 8
innodb_read_io_threads = 8
innodb_purge_threads = 8
innodb_page_cleaners = 8
innodb_open_files = 65535

innodb_buffer_pool_size=8G
```

#DISTINCT关键字经常在MySQL中使用，在MySQL5.7以前的版本中一般没有什么问题，但是在5.7以后的版本中会遇到这样的错误

```
sql_mode=NO_ENGINE_SUBSTITUTION,STRICT_TRANS_TABLES

max_connect_errors = 1000
#数据库隔离级别

transaction_isolation = READ-COMMITTED
#MySQL在完成某些join（连接）需求的时候，为了减少参与join的"被驱动表"的读取次数以提高性能

join_buffer_size = 128M
tmp_table_size = 64M

#该值设置过小将导致单个记录超过限制后写入数据库失败，且后续记录写入也将失败
max_allowed_packet = 130M
#MySQL在关闭一个交互的连接之前所要等待的秒数
interactive_timeout = 172800
#MySQL在关闭一个非交互的连接之前所要等待的秒数
wait_timeout = 172800
```

 配置文件里的路径配置是根据自己的MySQL安装目录来进行相应修改的。

（7）在命令窗口里执行命令mysqld --initialize-insecure --user=mysql，出现两个问题：一个是vs2013 c++，一个是关于动态链接库的，从网上下载msvcr120.dll和vcredist_x64.exe这两个文件，要注意操作系统版本和这两个文件版本一致，否则不能使用。

（8）将文件msvcr120.dll放到C:\Windows\system32下，文件vcredist_x64.exe直接运行后安装即可，然后再次运行第（7）步的操作命令。

（9）安装MySQL服务，输入"cmd"打开控制台，通过cd进入到MySQL目录，然后在控制台中输入"mysqld install"安装MySQL服务，如图1.7所示。

图1.7　安装MySQL服务

"cmd"需要用管理员权限打开（可以通过"开始"菜单→"所有程序"→"附件"→"命令提示符"→"以管理员身份运行"打开），否则可能会因为权限不够，而无法执行。

（10）启动服务，在控制台中键入"net start mysql"即可，启动MySQL服务如图1.8所示。

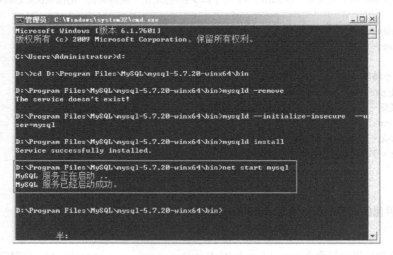

图1.8　启动MySQL服务

（11）启动之后，就可以登录了，输入mysql -u root -p（第一次登录没有密码），直接按回车键即可，如图1.9所示。

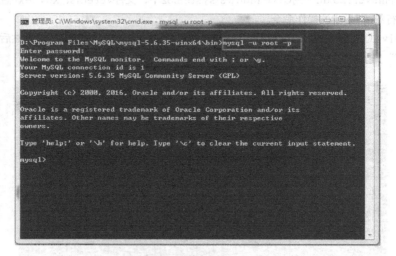

图1.9　登录MySQL

（12）更改root用户登录密码，以备日后使用。更改密码可以执行命令：UPDATE mysql.user SET authentication_string = PASSWORD('123456'), password_expired = 'N' WHERE User = 'root' AND Host = 'localhost';，如图1.10所示。

（13）MySQL新设置用户或更改密码后需用flush privileges刷新MySQL的系统权限相关表，否则会出现拒绝访问；还有一种方法，就是通过重新启动MySQL服务器来使新设置生效。

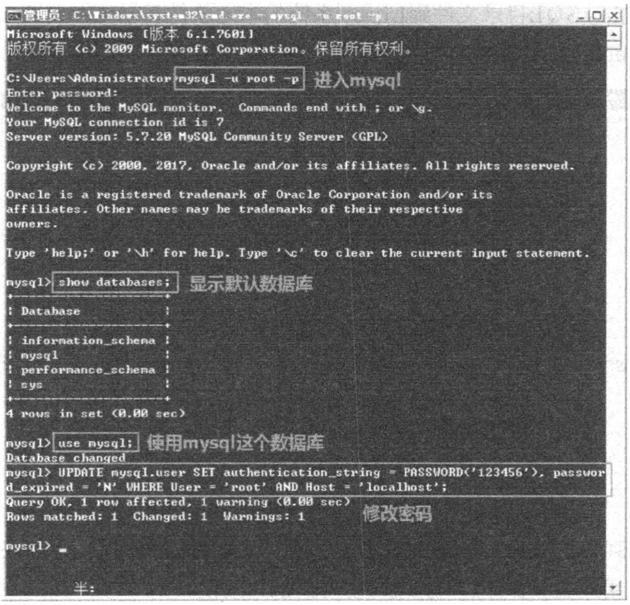

图1.10　更改root登录密码

1.5.2　Linux环境下安装

（1）在官网下载Linux版本，根据自己计算机的硬件情况下载合适的版本（如mysql-5.7.20-linux-glibc2.12-x86_64.tar.gz），如图1.11所示。

图1.11　Linux操作系统MySQL版本下载

（2）进入安装包所在目录（cd ..），解压MySQL压缩文件(tar -zxvf 文件)，解压安装文件，如图1.12所示。

图1.12　Linux解压安装文件

（3）复制解压后的MySQL目录文件到Linux系统的本地软件目录（/usr/local）中，执行命令：cp mysql-5.7.20-linux-glibc2.12-x86_64 -r；复制目录文件到系统本地，如图1.13所示。

图1.13 复制目录文件到系统本地

如果遇到"…Permission denied"，则说明权限不够，可以通过sudo权限的方式执行，如图1.14所示。

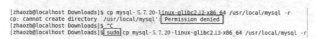

图1.14 sudo权限执行

（4）添加系统MySQL组（groupadd 组名），添加组如图1.15所示；为MySQL组添加MySQL用户（useradd -r -g mysql 用户名），添加MySQL用户如图1.16所示。

图1.15 添加MySQL组

图1.16 MySQL组添加MySQL用户

（5）进入MySQL安装目录（cd /user/local/mysql），修改当前目录拥有者为MySQL用户（chown -R mysql:mysql ./），如图1.17所示；再通过脚本初始化数据库（./scripts/mysql_install_db --user=mysql），如图1.18所示；修改当前目录下data目录拥有者为mysql（chown -R mysql:mysql data），用于存放数据库数据，如图1.19所示。

图1.17 修改当前拥有者权限 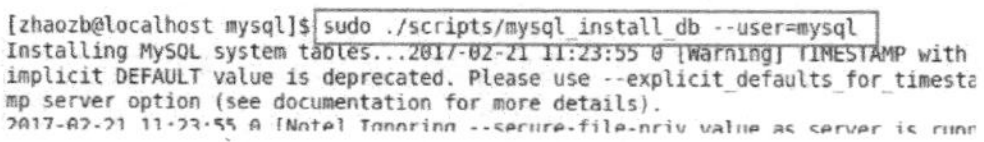　　　　　　　　图1.18 初始化表数据

（6）启动服务：service mysql start，MySQL就安装成功了，如图1.20所示。

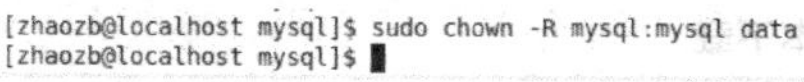

图1.19 修改data目录拥有者权限 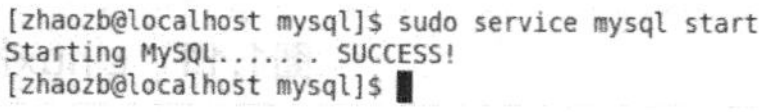　　　　　　　图1.20 启动MySQL服务

1.6 小结

本章介绍了MySQL的发展历史，即MySQL的诞生、发布以及现在的5.7.20最新版本，介绍了关系型数据库的概念和常用的关系型数据库，介绍了SQL结构和查询语言以及MySQL架构原理，对MySQL常用引擎InnoDB、MyISAM等进行了详细的讲解，最后讲解了在Windows/Linux环境下安装MySQL服务，为以后学习MySQL打下了基础。

第2章
MySQL数据类型

本章要点

- 数值类型
- 字符串类型
- 日期时间类型
- 复合类型
- 如何选取数据类型
- 小结

MySQL数据库是用来存储数据的，它可以存储姓名、性别、身高、年龄、出生日期、图片等不同的内容。本章介绍MySQL存储的数据类型。

存储数据内容可以使用不同的数据类型,如姓名可以存储成字符串类型、年龄可以存储成数值类型、出生日期可以存储成日期时间类型,如图2.1所示。MySQL常用的数据类型大致可以分为数值类型、字符串类型、日期时间类型、复合类型。

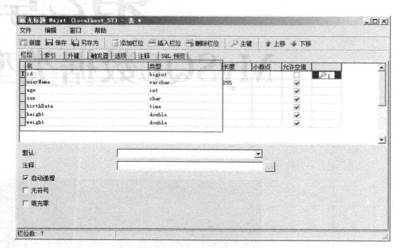

图2.1 数据类型存储

2.1 数值类型

MySQL支持所有标准SQL中的数值类型,包括严格数据类型(integer、smallint、decimal、numeric),以及近似数值数据类型(float、real、double),并在此基础上进行扩展。扩展后增加了tinyint、mediumint、bigint这3种长度不同的整型,并增加了bit位类型,用来存放位数据,如表2.1所示。

表2.1 数值类型

数值类型	字节	有符号存储范围	无符号存储范围	含义
tinyint	1字节	(−128, 127)	(0, 255)	小整数值
smallint	2字节	(−32768, 32 767)	(0, 65535)	大整数值
mediumint	3字节	(−8388608, 8388607)	(0, 16 777 215)	大整数值
int或integer	4字节	(−2 147 483 648, 2 147 483 647)	(0, 4 294 967 295)	大整数值
bigint	8字节	(−9233372036854775808, 9223372036854775 807)	(0, 18446744073709551615)	极大整数值
float(m,d)	4字节	−3.402823466E+38到−1.175494351E−38、0和1.175494351E−38到3.402823466E+38 m是数字总位数,d是小数点后面的位数,如果m和d被省略,根据硬件允许的限制来保存值		单精度浮点数值
double	8字节	−1.7976931348623157E+308到−2.225507385854072014E−308、0和2.225507385854072014E−308到1.7976931348623157E+308		双精度浮点数值
decimal(m,d)		m指定小数点左边和右边可以存储的十进制数字的最大个数,最大精度为38 d指定小数点右边可以存储的十进制数字的最大个数,小数位数必须是0~m的值,默认小数位数是0		定点数值

在int类型里，MySQL支持tinyint、smallint、mediumint、int、bigint 5种类型，这些类型都是用来存放整形数据的，只不过存储的值的大小范围不一样。

浮点类型MySQL支持float、double、decimal 3种类型，float是单精度浮点数值、double是双精度浮点数值、decimal是定点数值，如工资就可以使用float或者double类型来表示，如果确定数值总长度以及小数点位数，就可以用decimal定点数值来表示。如果将decimal定义成了decimal（5，2），最大的结果都是999.99，5是定点精度，2是小数位数。

存储数值范围越小，精度越高；存储数值范围越大，精度就越不准确。对于既要求精度又固定小数点位数的数值存储，采用decimal（numeric），其优点在于可以自定义小数点位数，精度高。某些特殊情况，如数值范围巨大则只能用float（real）类型了，此类型一般不提倡使用。

2.2 字符串类型

MySQL数据库提供了6种基本字符串类型：char、varchar、binary、varbinary、blob、text，它们用来存放一个字符、一个字符串、巨大的文本或者二进制字符串数据，如表2.2所示。

精讲视频

字符串类型

表2.2 字符串类型

数值类型	字 节	含 义
char	0~255字节	固定长度，最多255个字符
varchar	0~65 535字节	可变长度，最多65 535个字符
tinyblob	0~255字节	不超过255个字符的二进制字符串
tinytext	0~255字节	短文本字符串
blob	0~65 535字节	二进制形式的长文本数据
text	0~65 535字节	长文本数据
mediumblob	0~16 777 215字节	二进制形式的中等长度文本数据
mediumtext	0~16 777 215字节	中等长度文本数据
longblob	0~4 294 967 295字节	二进制形式的极大文本数据
longtext	0~4 294 967 295字节	极大文本数据
varbinary(m)	允许长度为0~m个字节的定长字节符串，值的长度+1个字节	
binary(m)	允许长度为0~m个字节的定长字节符串	

varchar可指定n，text不能指定，内部存储varchar是存入的实际字符数+1个字节（n<=255）或2个字节（n>255），text是实际字符数+2个字节；text类型不能有默认值；varchar可直接创建索引，text创建索引要指定前多少个字符。varchar的查询速度快于text，在都创建索引的情况下，text的索引似乎不起作用。

blob和text的存储方式不同，text以文本方式存储，英文存储区分大小写，而blob是以二进制方式存储，不分大小写；blob存储的数据只能整体读出；text可以指定字符集，blob不用指定字符集。

binary和varbinary类似于char和varchar，不同的是它们包含二进制字符串而不要非二进制字符串。也就是说，它们包含字节字符串而不是字符字符串。这说明它们没有字符集，并且排序和比较基于列值字节的数值。

2.3 日期时间类型

MySQL数据库提供了5种不同的日期时间类型：date、time、year、datetime、timestamp，用来存储和日期时间相关的数据，每个时间类型有一个有效值范围和一个"零"值，当指定不合法的MySQL不能表示的值时使用"零"值，如表2.3所示。

日期时间类型

表2.3 日期时间类型

数值类型	字节	范围	格式	含义
date	4字节	1000-01-01/9999-12-31	YYYY-MM-DD	日期值
time	3字节	'-838:59:59'/'838:59:59'	HH:MM:SS	时间值
year	1字节	1901/2155	YYYY	年份值
datetime	8字节	1000-01-01 00:00:00/9999-12-31 23:59:59	YYYY-MM-DD HH:MM:SS	混合日期和时间值
timestamp	4字节	1970-01-01 00:00:00/2037年某时	YYYYMMDD HHMMSS	混合日期和时间值，时间戳，自动存储记录修改时间

MySQL数据库用date日期类型和datetime日期时间类型来存储日期值，使用time时间类型来存储时间值，使用year来存放年份值，date 类型的值应该用连字号"-"分隔开，而 time 类型的值应该用冒号":"分隔开，若定义一个字段为timestamp，这个字段里的时间数据会在其他字段修改时自动刷新，所以这个数据类型的字段可以存放这条记录最后被修改的时间。

2.4 复合类型

MySQL数据库提供两种复合数据类型：enum单选字符串数据类型和set多选字符串数据类型。一个enum类型只允许从一个集合中取得一个值；而 set类型允许从一个集合中取得任意多个值。

复合类型

enum单选字符串数据类型适合存储表单界面中的"单选值"。设定enum的时候，需要给定"固定的几个选项"，存储的时候就只存储其中的一个值。设定enum的格式：enum("选项1","选项2","选项3",…)。实际上，enum的选项都会对应一个数字，依次是1，2，3，4，5…最多有65 535个选项，使用的时候，可以使用选项的字符串格式，也可以使用对应的数字。

set多选字符串数据类型适合存储表单界面中的"多选值"。设定set的时候，同样需要给定"固定的几个选项"，存储的时候可以存储其中的若干个值。设定set的格式：set("选项1","选项2","选项3",…)。同样地，set的每个选项值也对应一个数字，依次是1，2，4，8，16…最多有64个选项，使用的时候，可以使用set选项的字符串本身（多个选项用逗号分隔），也可以使用多个选项的数字之和（如3+7+8=18）。

2.5 如何选取数据类型

数据类型的选择会影响存储空间的开销和数据查询性能，所以在设计数据库的时候，应该为存储的数据选择正确的存储数据类型。

如何选取数据类型

可以遵循以下原则来选取数据类型。

（1）大小合适就是最好的，如存放姓名的字段类型，设置合适长度的字符串就可以，而不应该设置得非常大（65 535），因为这样可以用更少的磁盘容量、CPU缓存，大大减少了I/O开销。

（2）简单存放就好，如存放一个简单而又短小的字符串，可以选取varchar类型，而不应该选取text或者blob类型，简单的数据类型操作通常需要更少的CPU周期。

（3）尽量避免使用NULL，NULL是列默认的属性，通常我们要指定为NOT NULL。有NULL的列值会使得索引、索引统计和值比较更加复杂。

（4）数据类型的选择还应该考虑数据操作和应用处理的要求，如某些类型更适合于加快数据操作而不是其他方面。

（5）如果想用不同的标准对一列中所有的值进行排序，那么我们需要选择一种能够有效执行这些功能并且带有最大限度灵活性的类型。对相应字段添加索引，能够加快检索速度。需要保证使用的数据类型支持比较运算，并且相互兼容。

下面对比一下如何来选取数据类型。

1. char和varchar

char是固定长度的，查询速度比varchar快得多。char的缺点是浪费存储空间。检索char列时，返回的结果会删除尾部空格，所以程序需要对空格进行处理。

对于长度变化不大且对查询速度有较高要求的数据可以考虑使用char。随着MySQL的不断升级，varchar的性能不断改进并提高。

如果列中要存储的数据的长度差不多是一致的，则应该考虑使用char，否则应该考虑使用varchar；如果列中最大数据的长度小于50Byte，则一般也考虑使用char(当然，如果这个列很少用，则基于节省空间和减少I/O的考虑，还是可以使用varchar)；一般不宜定义大于50Byte的char类型列。

2. text和blob

在保存大文本时，通常选择text或者blob。二者的差别是blob可以保存二进制数据，如照片。text和blob又包括text、mediumtext、longtext和blob、mediumblob、longblob，它们之间的区别是存储文本长度不同和存储字节不同。在不必要的时候避免检索大型的blob或text值。把blob或text列分离到单独的表中。

3. 浮点数和定点数

浮点数存在误差问题，因此对于货币等对精度敏感的数据，应该用定点数表示或存储。在编程中，如果用到浮点数，要特别注意误差问题，并尽量避免做浮点数比较；要注意一些特殊值的处理。

decimal用于存储精确数据，而float只能用于存储非非精确的数据，故精确数据最好使用decimal类型；由于float的存储空间的开销一般比decimal小（精确到7位小数只需要4个字节，而精确到15位小数只需要8个字节），故非精确数据类型建议使用float。

4. 日期类型的选择

根据实际需要选择能够满足应用的最小存储日期类型。如果记录年、月、日、时、分、秒，并且记录年份比较久远，则最好使用datetime，不要使用timestamp；如果记录的日期需要让不同时区的用户使用，则最好使用timestamp，因为日期类型中只有它能够和实际时区相对应。

2.6 小结

本章主要讲解了MySQL数据类型，要掌握MySQL的4种数据类型：数值类型、字符串类型、日期时间类型、复合类型；要掌握各种数据类型里面有哪些存储方式，包括字节大小、范围，以及使用的场景；同时要学会如何选取数据类型，选择正确的数据类型对于数据库性能的影响很大，数据类型会影响存储空间的开销、数据的查询性能；要掌握各种类型存储的优、缺点以及适用的场景，这样才能设计出高效率的数据库。

第3章
MySQL 常用操作

本章要点

数据库用户管理
数据库操作
表操作
数据操作
字段操作
客户端操作数据库
小结

■ 本章采用命令行的方式来讲解MySQL的常用操作。通过本章，我们来学习如何管理数据库用户、连接MySQL服务、新增用户以及修改密码；创建数据库、修改数据库、删除数据库以及查看数据库操作；创建表、修改表结构、复制表以及临时表的使用；设置主键、设置复合主键、添加字段、改变字段类型、重命名、设置默认值以及自增字段的使用；插入数据、修改数据、删除数据、查询数据。除了使用命令行的方式来进行MySQL操作外，也可以使用客户端来进行操作，可以安装Navicat for MySQL客户端来进行MySQL的操作。

3.1 数据库用户管理

3.1.1 连接MySQL

精讲视频

数据库用户管理

连接MySQL包括两方面内容：一方面是连接本地MySQL；另一方面是连接远程MySQL。连接MySQL的命令格式如下。

mysql -h主机地址 -u用户名 -p用户密码

（1）连接本地MySQL。如果本地安装了MySQL数据库服务，用户名是root，密码是123456，可按如下步骤连接MySQL。

打开DOS命令行操作界面，进入到mysql/bin目录下面，输入命令"mysql -u root -p123456"，按回车键后就可以进入到MySQL中。如果刚安装好MySQL，超级用户root是没有密码的，故直接按回车键即可进入到MySQL中。

实战演练 ——数据库本地连接

Microsoft Windows [版本 6.1.7601]
版权所有 (c) 2009 Microsoft Corporation。保留所有权利。

C:\Users\Administrator>d:

D:\>cd D:\Program Files\MySQL\mysql-5.7.20-winx64\bin

D:\Program Files\MySQL\mysql-5.7.20-winx64\bin>mysql -u root -p123456
mysql: [Warning] Using a password on the command line interface can be insecure.

Welcome to the MySQL monitor. Commands end with ; or \g.
Your MySQL connection id is 19
Server version: 5.7.20 MySQL Community Server (GPL)

Copyright (c) 2000, 2017, Oracle and/or its affiliates. All rights reserved.

Oracle is a registered trademark of Oracle Corporation and/or its
affiliates. Other names may be trademarks of their respective
owners.

Type 'help;' or '\h' for help. Type '\c' to clear the current input statement.

mysql>

（2）连接远程MySQL。如果远程有一台MySQL服务器，IP是10.120.71.89，用户名是root，密码是123456，可按如下步骤连接MySQL。

实战演练 ——数据库远程连接

Microsoft Windows [版本 6.1.7601]
版权所有 (c) 2009 Microsoft Corporation。保留所有权利。

C:\Users\Administrator>mysql -h10.120.71.89 -u root -p123456
mysql: [Warning] Using a password on the command line interface can be insecure.

```
Welcome to the MySQL monitor.  Commands end with ; or \g.
Your MySQL connection id is 59237
Server version: 5.7.19-log MySQL Community Server (GPL)

Copyright (c) 2000, 2017, Oracle and/or its affiliates. All rights reserved.

Oracle is a registered trademark of Oracle Corporation and/or its
affiliates. Other names may be trademarks of their respective
owners.

Type 'help;' or '\h' for help. Type '\c' to clear the current input statement.

mysql>
```

 "-h"代表输入远程MySQL服务器的IP，"-u"代表输入用户名，"-p"代表输入密码；要退出MySQL服务，可以输入"quit"或者"exit"命令。

3.1.2 新增用户

MySQL数据库在安装的时候会创建一个管理员root用户，那么如果想再新增用户怎么办呢？新增用户的命令格式如下。

GRANT SELECT on 数据库.* to 用户名@登录主机 identified by "密码"

新增一个用户，用户名为shopdb，密码为shopdb_123456。让它可以在任何主机上登录，并对所有数据库有查询、插入、修改、删除的权限。

实战演练——新增用户

```
Microsoft Windows [版本 6.1.7601]
版权所有 (c) 2009 Microsoft Corporation。保留所有权利。

C:\Users\Administrator>mysql -u root -p123456
mysql: [Warning] Using a password on the command line interface can be insecure.

Welcome to the MySQL monitor.  Commands end with ; or \g.
Your MySQL connection id is 20
Server version: 5.7.20 MySQL Community Server (GPL)

Copyright (c) 2000, 2017, Oracle and/or its affiliates. All rights reserved.

Oracle is a registered trademark of Oracle Corporation and/or its
affiliates. Other names may be trademarks of their respective
owners.

Type 'help;' or '\h' for help. Type '\c' to clear the current input statement.

mysql> GRANT SELECT,INSERT,UPDATE,DELETE on *.* to shopdb@"%" identified by "sho
pdb_123456";
Query OK, 0 rows affected, 1 warning (0.71 sec)

mysql> exit;
Bye
```

```
C:\Users\Administrator>mysql –u shopdb –pshopdb_123456
mysql: [Warning] Using a password on the command line interface can be insecure.

Welcome to the MySQL monitor.  Commands end with ; or \g.
Your MySQL connection id is 21
Server version: 5.7.20 MySQL Community Server (GPL)

Copyright (c) 2000, 2017, Oracle and/or its affiliates. All rights reserved.

Oracle is a registered trademark of Oracle Corporation and/or its
affiliates. Other names may be trademarks of their respective
owners.

Type 'help;' or '\h' for help. Type '\c' to clear the current input statement.

mysql>
```

GRANT SELECT,INSERT,UPDATE,DELETE ON *.* to shopdb@"%" identified by "shopdb_123456";
新增用户命令里有"%"，代表Internet上的任何一台计算机都可以连接你的MySQL数据库，并进行相应的操作，这是很危险的一种授权方式。在实际项目中，用户将会被授予localhost本地访问或者指定服务器IP访问。

本地访问设置可以将"%"改成"localhost"，指定IP访问可以将"%"改成指定的IP地址，命令如下所示。

GRANT SELECT,INSERT,UPDATE,DELETE ON *.* to shopdb@localhost identified by "shopdb_123456";

GRANT SELECT,INSERT,UPDATE,DELETE ON *.* to shopdb@10.120.71.89 identified by "shopdb_123456";

3.1.3 修改用户密码

创建好的用户如果要修改密码，要更新MySQL的用户表，在5.7版本的MySQL中，密码存放在authentication_string中，需要对它进行修改。在修改密码的时候，用户需要获得reload权限，否则使用flush privileges刷新MySQL的系统权限相关表会报错，可以使用如下命令进行授权。

GRANT reload ON *.* to 'shopdb'@'%';

如果不对用户授予reload权限，也可以在更新密码之后重启MySQL服务，不使用flush privileges刷新MySQL的系统权限相关表。

实战演练——修改用户密码

```
Microsoft Windows [版本 6.1.7601]
版权所有 (c) 2009 Microsoft Corporation。保留所有权利。

C:\Users\Administrator>mysql –u root –p123456
mysql: [Warning] Using a password on the command line interface can be insecure.

Welcome to the MySQL monitor.  Commands end with ; or \g.
Your MySQL connection id is 28
Server version: 5.7.20 MySQL Community Server (GPL)

Copyright (c) 2000, 2017, Oracle and/or its affiliates. All rights reserved.

Oracle is a registered trademark of Oracle Corporation and/or its
affiliates. Other names may be trademarks of their respective
owners.
```

```
Type 'help;' or '\h' for help. Type '\c' to clear the current input statement.
#授权reload
mysql> GRANT reload ON *.* to 'shopdb'@'%';
Query OK, 0 rows affected (0.08 sec)

#修改密码
mysql> UPDATE mysql.user SET authentication_string = PASSWORD('123456'), password_expired = 'N' WHERE User = 'shopdb' AND Host = 'localhost';
Query OK, 0 rows affected, 1 warning (0.06 sec)
Rows matched: 0  Changed: 0  Warnings: 1

mysql> flush privileges;
Query OK, 0 rows affected (0.43 sec)

mysql>
```

> MySQL新设置用户或更改密码后，需用flush privileges刷新MySQL的系统权限相关表，否则会出现拒绝访问的情况。也可以重新启动MySQL服务器，来使新设置生效。

3.2 数据库操作

MySQL允许创建多个数据库，每个数据库承载不同的内容，使用命令操作可以查看数据库、创建数据库、使用数据库以及删除数据库。

精讲视频

数据库操作

3.2.1 查看数据库

使用SHOW databases命令可以查看有哪些数据库。

实战演练——查看数据库

```
Microsoft Windows [版本 6.1.7601]
版权所有 (c) 2009 Microsoft Corporation。保留所有权利。

C:\Users\Administrator>mysql –u root –p123456
mysql: [Warning] Using a password on the command line interface can be insecure.

Welcome to the MySQL monitor.  Commands end with ; or \g.
Your MySQL connection id is 32
Server version: 5.7.20 MySQL Community Server (GPL)

Copyright (c) 2000, 2017, Oracle and/or its affiliates. All rights reserved.

Oracle is a registered trademark of Oracle Corporation and/or its
affiliates. Other names may be trademarks of their respective
owners.

Type 'help;' or '\h' for help. Type '\c' to clear the current input statement.

mysql> SHOW databases;
```

```
+--------------------+
| Database           |
+--------------------+
| information_schema |
| mysql              |
| performance_schema |
| sys                |
| xjxt               |
+--------------------+
5 rows in set (0.00 sec)

mysql>
```

3.2.2 创建数据库

使用CREATE DATABASE databaseName命令可以创建数据库。

实战演练 —— 创建数据库

Microsoft Windows [版本 6.1.7601]
版权所有 (c) 2009 Microsoft Corporation。保留所有权利。

C:\Users\Administrator>mysql –u root –p123456
mysql: [Warning] Using a password on the command line interface can be insecure.
Welcome to the MySQL monitor. Commands end with ; or \g.
Your MySQL connection id is 34
Server version: 5.7.20 MySQL Community Server (GPL)

Copyright (c) 2000, 2017, Oracle and/or its affiliates. All rights reserved.

Oracle is a registered trademark of Oracle Corporation and/or its
affiliates. Other names may be trademarks of their respective
owners.

Type 'help;' or '\h' for help. Type '\c' to clear the current input statement.

```
mysql> CREATE DATABASE shop;
Query OK, 1 row affected (0.06 sec)

mysql> SHOW databases;
+--------------------+
| Database           |
+--------------------+
| information_schema |
| mysql              |
| performance_schema |
| shop               |
| sys                |
| xjxt               |
+--------------------+
```

```
6 rows in set (0.00 sec)

mysql>
```

3.2.3　使用数据库

要操作某个数据库或者数据库里的表，首先需选择要使用的数据库，使用 USE databaseName命令进入到数据库里。

实战演练——使用数据库

```
Microsoft Windows [版本 6.1.7601]
版权所有 (c) 2009 Microsoft Corporation。保留所有权利。

C:\Users\Administrator>mysql -u root -p123456
mysql: [Warning] Using a password on the command line interface can be insecure.

Welcome to the MySQL monitor.  Commands end with ; or \g.
Your MySQL connection id is 35
Server version: 5.7.20 MySQL Community Server (GPL)

Copyright (c) 2000, 2017, Oracle and/or its affiliates. All rights reserved.

Oracle is a registered trademark of Oracle Corporation and/or its
affiliates. Other names may be trademarks of their respective
owners.

Type 'help;' or '\h' for help. Type '\c' to clear the current input statement.

mysql> USE shop;
Database changed
mysql>
```

3.2.4　删除数据库

使用命令DROP DATABASE databaseName可以删除数据库。

实战演练——删除数据库

```
Microsoft Windows [版本 6.1.7601]
版权所有 (c) 2009 Microsoft Corporation。保留所有权利。

C:\Users\Administrator>mysql -u root -p123456
mysql: [Warning] Using a password on the command line interface can be insecure.

Welcome to the MySQL monitor.  Commands end with ; or \g.
Your MySQL connection id is 36
Server version: 5.7.20 MySQL Community Server (GPL)

Copyright (c) 2000, 2017, Oracle and/or its affiliates. All rights reserved.

Oracle is a registered trademark of Oracle Corporation and/or its
affiliates. Other names may be trademarks of their respective
owners.
```

Type 'help;' or '\h' for help. Type '\c' to clear the current input statement.

```
mysql> SHOW databases;
+--------------------+
| Database           |
+--------------------+
| information_schema |
| mysql              |
| performance_schema |
| shop               |
| sys                |
| xjxt               |
+--------------------+
6 rows in set (0.00 sec)

mysql> DROP DATABASE shop;
Query OK, 0 rows affected (0.46 sec)

mysql> SHOW databases;
+--------------------+
| Database           |
+--------------------+
| information_schema |
| mysql              |
| performance_schema |
| sys                |
| xjxt               |
+--------------------+
5 rows in set (0.06 sec)

mysql>
```

3.3 表操作

MySQL表操作是使用频率最高的操作。用户可以创建数据库表，查看数据库表，向数据库表里插入数据、更新数据、删除数据等，同时可以修改表结构、复制表、使用临时表。

3.3.1 创建表

创建一个shop数据库，在shop数据库里新建一个user用户表，包括用户id、姓名、性别、年龄、密码。

创建数据库表首先要进入到shop数据库里，然后使用create table user()命令来创建，括号里面是数据表的字段，包括用户（id）、姓名（name）、性别（sex）、年龄（age）、密码（password）。

设置表的用户id作为主键，自动递增并且不为空，同时设置字段的数据类型，存储引擎采用Innodb数据库引擎。

实战演练——创建表

Microsoft Windows [版本 6.1.7601]
版权所有 (c) 2009 Microsoft Corporation。保留所有权利。

```
C:\Users\Administrator>mysql -u root -p123456
mysql: [Warning] Using a password on the command line interface can be insecure.

Welcome to the MySQL monitor.  Commands end with ; or \g.
Your MySQL connection id is 37
Server version: 5.7.20 MySQL Community Server (GPL)

Copyright (c) 2000, 2017, Oracle and/or its affiliates. All rights reserved.

Oracle is a registered trademark of Oracle Corporation and/or its
affiliates. Other names may be trademarks of their respective
owners.

Type 'help;' or '\h' for help. Type '\c' to clear the current input statement.

mysql> CREATE DATABASE shop;
Query OK, 1 row affected (0.00 sec)

mysql> USE shop;
Database changed
mysql> CREATE TABLE user(
    -> id int(10) unsigned not null auto_increment,
    -> name varchar(25),
    -> sex varchar(5),
    -> age int(10),
    -> password varchar(25),
    -> primary key(id))engine=Innodb;
Query OK, 0 rows affected (0.90 sec)

mysql> SHOW tables;
+----------------+
| Tables_in_shop |
+----------------+
| user           |
+----------------+
1 row in set (0.04 sec)

mysql>
```

3.3.2 查看表结构

使用DESC tableName可以查看表结构。

实战演练 —— 查看表结构

```
Microsoft Windows [版本 6.1.7601]
版权所有 (c) 2009 Microsoft Corporation。保留所有权利。

C:\Users\Administrator>mysql -u root -p123456
mysql: [Warning] Using a password on the command line interface can be insecure.

Welcome to the MySQL monitor.  Commands end with ; or \g.
Your MySQL connection id is 38
Server version: 5.7.20 MySQL Community Server (GPL)
```

```
Copyright (c) 2000, 2017, Oracle and/or its affiliates. All rights reserved.

Oracle is a registered trademark of Oracle Corporation and/or its
affiliates. Other names may be trademarks of their respective
owners.

Type 'help;' or '\h' for help. Type '\c' to clear the current input statement.

mysql> USE shop;
Database changed
mysql> DESC user;
+----------+--------------+------+-----+---------+----------------+
| Field    | Type         | Null | Key | Default | Extra          |
+----------+--------------+------+-----+---------+----------------+
| id       | int(10) unsigned | NO | PRI | NULL    | auto_increment |
| name     | varchar(25)  | YES  |     | NULL    |                |
| sex      | varchar(5)   | YES  |     | NULL    |                |
| age      | int(10)      | YES  |     | NULL    |                |
| password | varchar(25)  | YES  |     | NULL    |                |
+----------+--------------+------+-----+---------+----------------+
5 rows in set (0.02 sec)

mysql>
```

3.3.3 复制表

MySQL可以快速复制表结构及数据，它以要复制表的结构和数据为基础，可以快速创建相同表结构和数据到新的表里，在开发过程中，可以复制一个新表作为测试表，而不用操作正式的表，以保证正在运行的数据不被破坏。复制表提供了两种方式：一种方式是可以复制表结构、数据、主键、索引；另一种方式是只能复制表结构、数据，不能复制主键和索引。

1. 第一种方式：复制表结构、数据、主键、索引

复制表结构、主键、索引，可执行如下命令。

CREATE TABLE new_table like old_table;

插入数据，可执行如下命令。

INSERT TABLE new_table SELECT * FROM old_table;

实战演练——复制表结构、数据、主键、索引

（1）基于user表的结构和数据进行复制，user表有主键和索引，如图3.1和图3.2所示。

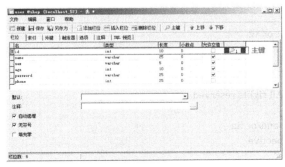

图3.1 user表主键

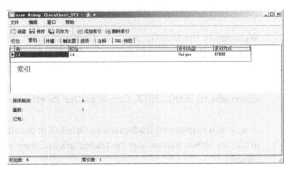

图3.2 user表索引

（2）在shop数据库的user表里添加一条记录，命令操作如下所示。

```
Microsoft Windows [版本 6.1.7601]
版权所有 (c) 2009 Microsoft Corporation。保留所有权利。

C:\Users\Administrator>mysql -u root -p123456
mysql: [Warning] Using a password on the command line interface can be insecure.

Welcome to the MySQL monitor.  Commands end with ; or \g.
Your MySQL connection id is 46
Server version: 5.7.20 MySQL Community Server (GPL)

Copyright (c) 2000, 2017, Oracle and/or its affiliates. All rights reserved.

Oracle is a registered trademark of Oracle Corporation and/or its
affiliates. Other names may be trademarks of their respective
owners.

Type 'help;' or '\h' for help. Type '\c' to clear the current input statement.

mysql> USE shop;
Database changed
mysql> INSERT INTO user VALUES(2,'tom','男','30','123456',0);
Query OK, 1 row affected (0.07 sec)

mysql> SELECT * FROM user;
+----+-------+-----+------+----------+-------+
| id | name  | sex | age  | password | phone |
+----+-------+-----+------+----------+-------+
|  1 | kevin | 男  |  20  | 123456   |     0 |
|  2 | tom   | 男  |  30  | 123456   |     0 |
+----+-------+-----+------+----------+-------+
2 rows in set (0.00 sec)

mysql>
```

（3）将user表的表结构、索引、主键复制到新的user_new表里，命令操作如下所示。

```
Microsoft Windows [版本 6.1.7601]
版权所有 (c) 2009 Microsoft Corporation。保留所有权利。

C:\Users\Administrator>mysql -u root -p123456
mysql: [Warning] Using a password on the command line interface can be insecure.

Welcome to the MySQL monitor.  Commands end with ; or \g.
Your MySQL connection id is 46
Server version: 5.7.20 MySQL Community Server (GPL)

Copyright (c) 2000, 2017, Oracle and/or its affiliates. All rights reserved.

Oracle is a registered trademark of Oracle Corporation and/or its
affiliates. Other names may be trademarks of their respective
owners.
```

```
Type 'help;' or '\h' for help. Type '\c' to clear the current input statement.

mysql> USE shop;
Database changed
mysql> INSERT INTO user VALUES(2,'tom','男','30','123456',0);
Query OK, 1 row affected (0.07 sec)

mysql> SELECT * FROM user;
+----+-------+-----+------+----------+-------+
| id | name  | sex | age  | password | phone |
+----+-------+-----+------+----------+-------+
|  1 | kevin | 男  |  20  | 123456   |     0 |
|  2 | tom   | 男  |  30  | 123456   |     0 |
+----+-------+-----+------+----------+-------+
2 rows in set (0.00 sec)

mysql> CREATE TABLE user_new like user;
Query OK, 0 rows affected (0.48 sec)

mysql> SELECT * FROM user_new;
Empty set (0.12 sec)

mysql>
```

（4）创建新的user_new表，只复制了表结构、主键和索引，并没有数据，把旧的表user的数据复制到新的表user_new，命令操作如下所示。

```
Microsoft Windows [版本 6.1.7601]
版权所有 (c) 2009 Microsoft Corporation。保留所有权利。

C:\Users\Administrator>mysql -u root -p123456
mysql: [Warning] Using a password on the command line interface can be insecure.

Welcome to the MySQL monitor.  Commands end with ; or \g.
Your MySQL connection id is 46
Server version: 5.7.20 MySQL Community Server (GPL)

Copyright (c) 2000, 2017, Oracle and/or its affiliates. All rights reserved.

Oracle is a registered trademark of Oracle Corporation and/or its
affiliates. Other names may be trademarks of their respective
owners.

Type 'help;' or '\h' for help. Type '\c' to clear the current input statement.

mysql> INSERT INTO user VALUES(2,'tom','男','30','123456',0);
ERROR 1046 (3D000): No database SELECTed
mysql> USE shop;
Database changed
mysql> INSERT INTO user VALUES(2,'tom','男','30','123456',0);
Query OK, 1 row affected (0.07 sec)

mysql> SELECT * FROM user;
+----+-------+-----+------+----------+-------+
```

```
+----+-------+-----+-----+----------+-------+
| id | name  | sex | age | password | phone |
+----+-------+-----+-----+----------+-------+
|  1 | kevin | 男  |  20 | 123456   |     0 |
|  2 | tom   | 男  |  30 | 123456   |     0 |
+----+-------+-----+-----+----------+-------+
2 rows in set (0.00 sec)

mysql> CREATE TABLE user_new like user;
Query OK, 0 rows affected (0.48 sec)

mysql> SELECT * FROM user_new;
Empty set (0.12 sec)

mysql> INSERT user_new SELECT * FROM user;
Query OK, 2 rows affected (0.13 sec)
Records: 2  Duplicates: 0  Warnings: 0

mysql> SELECT * FROM user_new;
+----+-------+-----+-----+----------+-------+
| id | name  | sex | age | password | phone |
+----+-------+-----+-----+----------+-------+
|  1 | kevin | 男  |  20 | 123456   |     0 |
|  2 | tom   | 男  |  30 | 123456   |     0 |
+----+-------+-----+-----+----------+-------+
2 rows in set (0.00 sec)

mysql>
```

2. 第二种方式：复制表结构、数据，不能复制主键、索引

复制表结构、数据，可执行如下命令。

CREATE TABLE new_table SELECT * FROM old_table;

复制表结构，不复制数据，可执行如下命令。

CREATE TABLE new_table SELECT * FROM old_table WHERE 0;

实战演练——复制表结构、数据

（1）将user表的表结构、索引、主键复制到新的user_new2表里，可以查看到user_new2表里已经复制进来数据了，命令操作如下所示。

```
Microsoft Windows [版本 6.1.7601]
版权所有 (c) 2009 Microsoft Corporation。保留所有权利。

C:\Users\Administrator>mysql -u root -p123456
mysql: [Warning] Using a password on the command line interface can be insecure.

Welcome to the MySQL monitor.  Commands end with ; or \g.
Your MySQL connection id is 48
Server version: 5.7.20 MySQL Community Server (GPL)

Copyright (c) 2000, 2017, Oracle and/or its affiliates. All rights reserved.

Oracle is a registered trademark of Oracle Corporation and/or its
affiliates. Other names may be trademarks of their respective
```

owners.

Type 'help;' or '\h' for help. Type '\c' to clear the current input statement.

```
mysql> USE shop;
Database changed
mysql> CREATE TABLE user_new2 SELECT * FROM user;
Query OK, 2 rows affected (0.68 sec)
Records: 2  Duplicates: 0  Warnings: 0

mysql> SELECT * FROM user_new2;
+----+-------+-----+------+----------+-------+
| id | name  | sex | age  | password | phone |
+----+-------+-----+------+----------+-------+
|  1 | kevin | 男  |  20  | 123456   |     0 |
|  2 | tom   | 男  |  30  | 123456   |     0 |
+----+-------+-----+------+----------+-------+
2 rows in set (0.00 sec)

mysql>
```

（2）将user表的表结构、索引、主键复制到新的user_new3表里，可以查看到user_new3表里没有数据，命令操作如下所示。

```
Microsoft Windows [版本 6.1.7601]
版权所有 (c) 2009 Microsoft Corporation。保留所有权利。

C:\Users\Administrator>mysql -u root -p123456
mysql: [Warning] Using a password on the command line interface can be insecure.

Welcome to the MySQL monitor.  Commands end with ; or \g.
Your MySQL connection id is 50
Server version: 5.7.20 MySQL Community Server (GPL)

Copyright (c) 2000, 2017, Oracle and/or its affiliates. All rights reserved.

Oracle is a registered trademark of Oracle Corporation and/or its
affiliates. Other names may be trademarks of their respective
owners.

Type 'help;' or '\h' for help. Type '\c' to clear the current input statement.

mysql> USE shop;
Database changed
mysql> CREATE TABLE user_new3 SELECT * FROM user WHERE 0;
Query OK, 0 rows affected (0.67 sec)
Records: 0  Duplicates: 0  Warnings: 0

mysql> SELECT * FROM user_new3;
Empty set (0.00 sec)

mysql>
```

3.3.4 临时表和内存表

MySQL临时表主要用于对大数据量表做一个临时表，以提高查询速度。临时表建在内存里，数据在内存里，缺省存储引擎为MySQL服务器默认引擎，引擎类型只能是MEMORY（HEAP）、MyISAM、MERGE、InnoDB。

MySQL内存表也可以对大数据量表做一个临时表，以提高查询速度，会把表结构存放在磁盘上，把数据放在内存中，缺省存储引擎为MEMORY。

创建临时表，命令执行如下所示。

CREATE temporary TABLE tmp1(id int not null);

创建内存表，命令执行如下所示。

CREATE TABLE tmp2(id int not null) ENGINE=MEMORY;

实战演练——临时表和内存表

```
Microsoft Windows [版本 6.1.7601]
版权所有 (c) 2009 Microsoft Corporation。保留所有权利。

C:\Users\Administrator>mysql -u root -p123456
mysql: [Warning] Using a password on the command line interface can be insecure.

Welcome to the MySQL monitor.  Commands end with ; or \g.
Your MySQL connection id is 51
Server version: 5.7.20 MySQL Community Server (GPL)

Copyright (c) 2000, 2017, Oracle and/or its affiliates. All rights reserved.

Oracle is a registered trademark of Oracle Corporation and/or its
affiliates. Other names may be trademarks of their respective
owners.

Type 'help;' or '\h' for help. Type '\c' to clear the current input statement.

mysql> USE shop;
Database changed
mysql> CREATE temporary TABLE tmp1(id int not null);
Query OK, 0 rows affected (0.26 sec)

mysql> SHOW CREATE TABLE tmp1;
+-------+------------------------------------------------------------+
| Table | Create Table                                               |
+-------+------------------------------------------------------------+
| tmp1  | CREATE TEMPORARY TABLE `tmp1` ( `id` int(11) NOT NULL
) ENGINE=InnoDB DEFAULT CHARSET=utf8                                 |
+-------+------------------------------------------------------------+
1 row in set (0.00 sec)

mysql> CREATE TABLE tmp2(id int not null) ENGINE=MEMORY;
Query OK, 0 rows affected (0.19 sec)

mysql> SHOW CREATE TABLE tmp2
```

```
+-------+----------------------------------------+
| Table | Create Table                           |
+-------+----------------------------------------+
| tmp2  | CREATE TABLE 'tmp2' (  'id' int(11) NOT NULL
) ENGINE=MEMORY DEFAULT CHARSET=utf8 |
+-------+----------------------------------------+
1 row in set (0.15 sec)

mysql>
```

临时表和内存表的区别如下。

（1）临时表的表结构和数据都保存在内存里；内存表的表结构保存在磁盘里，数据和索引保存在内存里。

（2）临时表使用的缺省存储引擎为MySQL服务器默认引擎；内存表使用的存储引擎为MEMORY服务器引擎。

（3）临时表可以通过参数 tmp_table_size 来设定临时表的大小；内存表可以通过参数max_heap_table_size来设定内存表的大小。

（4）临时表到达tmp_table_size设定的内存上限后将在磁盘上创建临时文件；内存表到达max_heap_table_size设定的内存上限后将报错。

（5）临时表可以包含TEXT、BLOB等字段；内存表不能包含TEXT、BLOB等字段。

（6）临时表一般比较少用，通常是在应用程序中动态创建或者由MySQL内部根据SQL执行计划自己创建；内存表则大多作为Cache来使用，特别在没有第三方Cache使用时，随着memcache、NoSQL的流行，越来越少选择使用内存表。

（7）临时表只在当前连接可见，当关闭连接时，MySQL会自动删除表并释放所有空间；内存表在MySQL重启后，主键、自增、索引仍然存在，只是数据丢失。

临时表不能使用rename来重命名，但是可以用alter table rename来代替；可以复制临时表得到一个新的临时表，格式为CREATE temporary table new_table SELECT * FROM old_table；在同一个查询语句中，相同的临时表只能出现一次，但不同的临时表可以出现在同一个查询语句中。

3.4 数据操作

插入、查询、修改和删除是MySQL数据库的4种最基本的操作，在项目开发的过程中也使用得最频繁。在MySQL里，插入使用INSERT关键字、查询使用SELECT关键字、修改使用UPDATE关键字、删除使用DELETE关键字。

数据操作

3.4.1 插入（INSERT）、查询（SELECT）

将一条数据插入到数据库里，可以使用如下命令。

INSERT INTO 表名(字段名, 字段名) VALUES(值, 值);

或者可以不指明表里的字段，但是值按字段的顺序插入，可以使用如下命令。

INSERT INTO 表名 VALUES(值, 值);

查询数据，可以使用如下命令。

SELECT * FROM 表名;

或者查询指定字段,可以使用如下命令。

```
SELECT id FROM 表名;
```
使用WHERE条件语句按条件查询,把某一列或者几列作为查询条件,可以使用如下命令。
```
SELECT * FROM 表名 WHERE id=10;

SELECT * FROM 表名 WHERE id=10 and name='小明';
```

实战演练——数据插入和查询

Microsoft Windows [版本 6.1.7601]
版权所有 (c) 2009 Microsoft Corporation。保留所有权利。

#进入mysql数据库
```
C:\Users\Administrator>mysql –u root –p123456
mysql: [Warning] Using a password on the command line interface can be insecure.

Welcome to the MySQL monitor.  Commands end with ; or \g.
Your MySQL connection id is 11
Server version: 5.7.20 MySQL Community Server (GPL)

Copyright (c) 2000, 2017, Oracle and/or its affiliates. All rights reserved.

Oracle is a registered trademark of Oracle Corporation and/or its
affiliates. Other names may be trademarks of their respective
owners.

Type 'help;' or '\h' for help. Type '\c' to clear the current input statement.
```

#使用shop数据库
```
mysql> USE shop;
Database changed
```

#查询user表结构
```
mysql> DESC user;
+----------+------------------+------+-----+---------+----------------+
| Field    | Type             | Null | Key | Default | Extra          |
+----------+------------------+------+-----+---------+----------------+
| id       | int(10) unsigned | NO   | PRI | NULL    | auto_increment |
| name     | varchar(25)      | YES  |     | NULL    |                |
| sex      | varchar(5)       | YES  |     | NULL    |                |
| age      | int(10)          | YES  |     | NULL    |                |
| password | varchar(25)      | YES  |     | NULL    |                |
| phone    | int(25)          | YES  |     | NULL    |                |
+----------+------------------+------+-----+---------+----------------+
6 rows in set (0.03 sec)
```

#插入数据,指定列名
```
mysql> INSERT INTO user(id,name,sex,age,password,phone) VALUES(3,'david','男','28','111111',null);
Query OK, 1 row affected (1.59 sec)
```

#插入数据,不指定列名
```
mysql> INSERT INTO user VALUES(4,'lili','女','25','222222','0');
Query OK, 1 row affected (0.00 sec)
```

```
#查询所有字段语句
mysql> SELECT * FROM user;
+----+-------+-----+-----+----------+-------+
| id | name  | sex | age | password | phone |
+----+-------+-----+-----+----------+-------+
|  1 | kevin | 男  |  20 | 123456   |     0 |
|  2 | tom   | 男  |  30 | 123456   |     0 |
|  3 | david | 男  |  28 | 111111   |  NULL |
|  4 | lili  | 女  |  25 | 222222   |     0 |
+----+-------+-----+-----+----------+-------+
4 rows in set (0.00 sec)

#查询指定字段语句
mysql> SELECT name FROM user;
+-------+
| name  |
+-------+
| kevin |
| tom   |
| david |
| lili  |
+-------+
4 rows in set (0.00 sec)

#查询性别为男的数据
mysql> SELECT * FROM user WHERE sex='男';
+----+-------+-----+-----+----------+-------+
| id | name  | sex | age | password | phone |
+----+-------+-----+-----+----------+-------+
|  1 | kevin | 男  |  20 | 123456   |     0 |
|  2 | tom   | 男  |  30 | 123456   |     0 |
|  3 | david | 男  |  28 | 111111   |  NULL |
+----+-------+-----+-----+----------+-------+
3 rows in set (0.00 sec)

#查询性别为男并且名字是david的数据
mysql> SELECT * FROM user WHERE sex='男' and name='david';
+----+-------+-----+-----+----------+-------+
| id | name  | sex | age | password | phone |
+----+-------+-----+-----+----------+-------+
|  1 | david | 男  |  20 | 123456   |     0 |
+----+-------+-----+-----+----------+-------+
1 row in set (0.00 sec)

mysql>
```

3.4.2 修改记录（UPDATE）

在MySQL里修改使用关键字UPDATE，命令如下所示。

UPDATE 表名 SET 字段=值，字段=值 WHERE 条件

UPDATE user SET name='小明',sex='男' WHERE id = 4;

实战演练——修改记录

Microsoft Windows [版本 6.1.7601]
版权所有 (c) 2009 Microsoft Corporation。保留所有权利。
#输入用户名密码，进入数据库
C:\Users\Administrator>mysql -u root -p123456
mysql: [Warning] Using a password on the command line interface can be insecure.

Welcome to the MySQL monitor. Commands end with ; or \g.
Your MySQL connection id is 13
Server version: 5.7.20 MySQL Community Server (GPL)

Copyright (c) 2000, 2017, Oracle and/or its affiliates. All rights reserved.

Oracle is a registered trademark of Oracle Corporation and/or its
affiliates. Other names may be trademarks of their respective
owners.

Type 'help;' or '\h' for help. Type '\c' to clear the current input statement.

#使用shop数据库
mysql> USE shop;
Database changed

#查询user表数据
mysql> SELECT * FROM user;

```
+----+-------+------+------+----------+-------+
| id | name  | sex  | age  | password | phone |
+----+-------+------+------+----------+-------+
|  1 | kevin | 男   |   20 | 123456   |     0 |
|  2 | tom   | 男   |   30 | 123456   |     0 |
|  3 | david | 男   |   28 | 111111   |  NULL |
|  4 | lili  | 女   |   25 | 222222   |     0 |
+----+-------+------+------+----------+-------+
```
4 rows in set (0.00 sec)

#修改一个字段的值，将david性别设置为女
mysql> UPDATE user SET sex='女' WHERE name='david';
Query OK, 1 row affected (0.00 sec)
Rows matched: 1 Changed: 1 Warnings: 0

#查询user表数据
mysql> SELECT * FROM user;

```
+----+-------+------+------+----------+-------+
| id | name  | sex  | age  | password | phone |
+----+-------+------+------+----------+-------+
|  1 | kevin | 男   |   20 | 123456   |     0 |
|  2 | tom   | 男   |   30 | 123456   |     0 |
|  3 | david | 女   |   28 | 111111   |  NULL |
|  4 | lili  | 女   |   25 | 222222   |     0 |
+----+-------+------+------+----------+-------+
```

4 rows in set (0.00 sec)

#修改多个字段的值，将id等于4的这一行的名字改为小明，性别改为男
mysql> UPDATE user SET name='小明',sex='男' WHERE id=4;
Query OK, 1 row affected (0.00 sec)
Rows matched: 1 Changed: 1 Warnings: 0

#查收user表数据
mysql> SELECT * FROM user;
+----+-------+-----+------+----------+-------+
| id | name | sex | age | password | phone |
+----+-------+-----+------+----------+-------+
1	kevin	男	20	123456	0
2	tom	男	30	123456	0
3	david	女	28	111111	NULL
4	小明	男	25	222222	0
+----+-------+-----+------+----------+-------+
4 rows in set (0.00 sec)

mysql>

3.4.3 删除记录（DELETE）

在MySQL中删除数据可以使用DELETE关键字，命令如下所示。

DELETE FROM 表名 WHERE 条件

DELETE FROM user WHERE id=4;

实战演练 ——删除记录

Microsoft Windows [版本 6.1.7601]
版权所有 (c) 2009 Microsoft Corporation。保留所有权利。

#输入用户名密码，进入mysql数据库
C:\Users\Administrator>mysql –u root –p123456
mysql: [Warning] Using a password on the command line interface can be insecure.

Welcome to the MySQL monitor. Commands end with ; or \g.
Your MySQL connection id is 15
Server version: 5.7.20 MySQL Community Server (GPL)

Copyright (c) 2000, 2017, Oracle and/or its affiliates. All rights reserved.

Oracle is a registered trademark of Oracle Corporation and/or its
affiliates. Other names may be trademarks of their respective
owners.

Type 'help;' or '\h' for help. Type '\c' to clear the current input statement.

#使用shop数据库
mysql> USE shop;
Database changed

```
#查询user表数据
mysql> SELECT * FROM user;
+----+-------+-----+-----+----------+-------+
| id | name  | sex | age | password | phone |
+----+-------+-----+-----+----------+-------+
|  1 | kevin | 男  |  20 | 123456   |     0 |
|  2 | tom   | 男  |  30 | 123456   |     0 |
|  3 | david | 女  |  28 | 111111   |  NULL |
|  4 | 小明  | 男  |  25 | 222222   |     0 |
+----+-------+-----+-----+----------+-------+
4 rows in set (0.00 sec)

#删除id等于4的数据
mysql> DELETE FROM user WHERE id=4;
Query OK, 1 row affected (0.00 sec)

#查询user表数据
mysql> SELECT * FROM user;
+----+-------+-----+-----+----------+-------+
| id | name  | sex | age | password | phone |
+----+-------+-----+-----+----------+-------+
|  1 | kevin | 男  |  20 | 123456   |     0 |
|  2 | tom   | 男  |  30 | 123456   |     0 |
|  3 | david | 女  |  28 | 111111   |  NULL |
+----+-------+-----+-----+----------+-------+
3 rows in set (0.00 sec)

#删除性别为男的数据
mysql> DELETE FROM user WHERE sex='男';
Query OK, 2 rows affected (0.00 sec)

#查询user表数据
mysql> SELECT * FROM user;
+----+-------+-----+-----+----------+-------+
| id | name  | sex | age | password | phone |
+----+-------+-----+-----+----------+-------+
|  3 | david | 女  |  28 | 111111   |  NULL |
+----+-------+-----+-----+----------+-------+
1 row in set (0.00 sec)

mysql>
```

3.4.4 对查询结果排序（ORDER BY）

在MySQL数据库中，使用ORDER BY进行排序，使用关键字ASC进行升序排序，使用关键字DESC进行降序排序，同时可以按一个字段或者多个字段进行排序。如果按多个字段进行排序，先进行第一个字段的排序，然后在结果集里面再进行第二个字段的排序，以此类推。

（1）ORDER BY column ASC：按某一字段进行升序排序，ASC可以省略不写。

SELECT * FROM user ORDER BY id ASC;
或者
SELECT * FROM user ORDER BY id;

（2）ORDER BY column DESC：按某一字段进行降序排序，DESC不可以省略不写。
SELECT * FROM user ORDER BY id DESC;
（3）ORDER BY column1,column2 DESC：按多个字段进行降序排序。
SELECT * FROM user ORDER BY sex,age DESC;

实战演练——查询结果排序

Microsoft Windows [版本 6.1.7601]
版权所有 (c) 2009 Microsoft Corporation。保留所有权利。

C:\Users\Administrator>mysql –u root –p123456
mysql: [Warning] Using a password on the command line interface can be insecure.

Welcome to the MySQL monitor. Commands end with ; or \g.
Your MySQL connection id is 27
Server version: 5.7.20 MySQL Community Server (GPL)

Copyright (c) 2000, 2017, Oracle and/or its affiliates. All rights reserved.

Oracle is a registered trademark of Oracle Corporation and/or its
affiliates. Other names may be trademarks of their respective
owners.

Type 'help;' or '\h' for help. Type '\c' to clear the current input statement.

mysql> USE shop;
Database changed

#查询user表所有数据
mysql> SELECT * FROM user;
+----+------+------+------+----------+-------+
| id | name | sex | age | password | phone |
+----+------+------+------+----------+-------+
3	david	女	28	111111	NULL
4	小红	女	27	123456	NULL
5	小明	男	10	123456	NULL
6	小刚	男	12	123456	NULL
7	小王	男	14	111111	NULL
8	小绿	女	34	222222	NULL
9	晓峰	男	15	333333	NULL
10	小影	女	26	444444	NULL
11	大梅	女	27	555555	NULL
+----+------+------+------+----------+-------+
9 rows in set (0.00 sec)

#按id升序排序查询
mysql> SELECT * FROM user ORDER BY id ASC;
+----+------+------+------+----------+-------+
| id | name | sex | age | password | phone |
+----+------+------+------+----------+-------+
| 3 | david| 女 | 28 | 111111 | NULL |

```
|  4 | 小红  | 女  | 27 | 123456  | NULL |
|  5 | 小明  | 男  | 10 | 123456  | NULL |
|  6 | 小刚  | 男  | 12 | 123456  | NULL |
|  7 | 小王  | 男  | 14 | 111111  | NULL |
|  8 | 小绿  | 女  | 34 | 222222  | NULL |
|  9 | 晓峰  | 男  | 15 | 333333  | NULL |
| 10 | 小影  | 女  | 26 | 444444  | NULL |
| 11 | 大梅  | 女  | 27 | 555555  | NULL |
+----+------+-----+----+---------+------+
9 rows in set (0.00 sec)
```

#按id降序排序查询
```
mysql> SELECT * FROM user ORDER BY id DESC;
+----+------+-----+----+---------+------+
| id | name | sex | age| password| phone|
+----+------+-----+----+---------+------+
| 11 | 大梅  | 女  | 27 | 555555  | NULL |
| 10 | 小影  | 女  | 26 | 444444  | NULL |
|  9 | 晓峰  | 男  | 15 | 333333  | NULL |
|  8 | 小绿  | 女  | 34 | 222222  | NULL |
|  7 | 小王  | 男  | 14 | 111111  | NULL |
|  6 | 小刚  | 男  | 12 | 123456  | NULL |
|  5 | 小明  | 男  | 10 | 123456  | NULL |
|  4 | 小红  | 女  | 27 | 123456  | NULL |
|  3 | david| 女  | 28 | 111111  | NULL |
+----+------+-----+----+---------+------+
9 rows in set (0.00 sec)
```

#先按性别查询，再按年龄升序排序查询
```
mysql> SELECT * FROM user ORDER BY sex DESC,age ASC;
+----+------+-----+----+---------+------+
| id | name | sex | age| password| phone|
+----+------+-----+----+---------+------+
|  5 | 小明  | 男  | 10 | 123456  | NULL |
|  6 | 小刚  | 男  | 12 | 123456  | NULL |
|  7 | 小王  | 男  | 14 | 111111  | NULL |
|  9 | 晓峰  | 男  | 15 | 333333  | NULL |
| 10 | 小影  | 女  | 26 | 444444  | NULL |
|  4 | 小红  | 女  | 27 | 123456  | NULL |
| 11 | 大梅  | 女  | 27 | 555555  | NULL |
|  3 | david| 女  | 28 | 111111  | NULL |
|  8 | 小绿  | 女  | 34 | 222222  | NULL |
+----+------+-----+----+---------+------+
9 rows in set (0.00 sec)

mysql>
```

3.4.5 对查询结果分组（GROUP BY）

GROUP BY对查询结果分组是将查询结果按照1个或多个字段进行分组，字段值相同的为一组，GROUP BY可以用于单个字段和多个字段。

```
SELECT * FROM user GROUP BY sex;
```

group_concat（字段名）可以作为一个输出字段来使用，表示分组之后，根据分组结果，使用group_concat（）来放置每一组的某字段的值的集合。

SELECT sex,group_concat(name) FROM user GROUP BY sex;

实战演练 ——查询结果分组

下面按用户性别进行分组查询，使用group_concat（name）输出用户的姓名。

Microsoft Windows [版本 6.1.7601]
版权所有 (c) 2009 Microsoft Corporation。保留所有权利。

C:\Users\Administrator>mysql -u root -p123456
mysql: [Warning] Using a password on the command line interface can be insecure.

Welcome to the MySQL monitor. Commands end with ; or \g.
Your MySQL connection id is 549
Server version: 5.7.20 MySQL Community Server (GPL)

Copyright (c) 2000, 2017, Oracle and/or its affiliates. All rights reserved.

Oracle is a registered trademark of Oracle Corporation and/or its
affiliates. Other names may be trademarks of their respective
owners.

Type 'help;' or '\h' for help. Type '\c' to clear the current input statement.

mysql> USE shop;
Database changed

mysql> SELECT * FROM user;
+----+------+-----+-----+--------+-------+-----------+--------+
| id | name | sex | age | password | phone | loginName | remark |
+----+------+-----+-----+--------+-------+-----------+--------+
3	david	女	28	111111	NULL	david	baann
4	小红	女	27	123456	NULL	xiaohong	black
5	小明	男	10	123456	NULL	xiaoming	berry
6	小刚	男	12	123456	NULL	xiaogang	banner
7	小王	男	14	111111	NULL	xiaowang	banana
8	小绿	女	34	222222	NULL	xiaolv	car
9	晓峰	男	15	333333	NULL	xiaofeng	carray
10	小影	女	26	444444	NULL	xiaoying	baaaa
11	大梅	女	27	555555	NULL	damei	accc
+----+------+-----+-----+--------+-------+-----------+--------+
9 rows in set (0.00 sec)

#根据用户性别进行分组，性别sex字段的全部值只有两个('男'和'女')，所以分为了两组，当GROUP BY单独使用时，只显示出每组的第一条记录

mysql> SELECT * FROM user GROUP BY sex;
+----+------+-----+-----+--------+-------+-----------+--------+
| id | name | sex | age | password | phone | loginName | remark |
+----+------+-----+-----+--------+-------+-----------+--------+

```
| 3 | david | 女 | 28 | 111111 | NULL | david   | baann |
| 5 | 小明  | 男 | 10 | 123456 | NULL | xiaoming| berry |
+---+-------+----+----+--------+------+---------+-------+
2 rows in set (0.00 sec)
```

#根据用户性别和id进行分组,因为id是没有重复的,所以按性别把女和男分开显示
```
mysql> SELECT * FROM user GROUP BY sex,id;
+----+-------+-----+-----+----------+------+----------+--------+
| id | name  | sex | age | password | phone| loginName| remark |
+----+-------+-----+-----+----------+------+----------+--------+
| 3  | david | 女  | 28  | 111111   | NULL | david    | baann  |
| 4  | 小红  | 女  | 27  | 123456   | NULL | xiaohong | black  |
| 8  | 小绿  | 女  | 34  | 222222   | NULL | xiaolv   | car    |
| 10 | 小影  | 女  | 26  | 444444   | NULL | xiaoying | baaaa  |
| 11 | 大梅  | 女  | 27  | 555555   | NULL | damei    | accc   |
| 5  | 小明  | 男  | 10  | 123456   | NULL | xiaoming | berry  |
| 6  | 小刚  | 男  | 12  | 123456   | NULL | xiaogang | banner |
| 7  | 小王  | 男  | 14  | 111111   | NULL | xiaowang | banana |
| 9  | 晓峰  | 男  | 15  | 333333   | NULL | xiaofeng | carray |
+----+-------+-----+-----+----------+------+----------+--------+
9 rows in set (0.00 sec)
```

#按性别分组,输出用户的姓名
```
mysql> SELECT sex,group_concat(name) FROM user GROUP BY sex;
+-----+-----------------------------+
| sex | group_concat(name)          |
+-----+-----------------------------+
| 女  | david,小红,小绿,小影,大梅   |
| 男  | 小明,小刚,小王,晓峰         |
+-----+-----------------------------+
2 rows in set (0.06 sec)
```

#按性别分组,输出用户的id
```
mysql> SELECT sex,group_concat(id) FROM user GROUP BY sex;
+-----+------------------+
| sex | group_concat(id) |
+-----+------------------+
| 女  | 3,4,8,10,11      |
| 男  | 5,6,7,9          |
+-----+------------------+
2 rows in set (0.00 sec)

mysql>
```

3.4.6 设置分组条件(HAVING)

HAVING 是用来设置分组条件的条件表达式,用来在分组查询后指定一些条件来输出查询结果,WHERE语句在聚合前先筛选记录,也就是说作用在GROUP BY和HAVING子句前,而 HAVING子句在聚合后对组记录进行筛选,HAVING只能用于GROUP BY。

```
SELECT sex,count(sex) FROM user WHERE age > 15  GROUP BY sex HAVING count(sex)>2;
```

实战演练——设置分组条件

下面查询按性别分组，然后查询在性别分组后大于4的性别。

Microsoft Windows [版本 6.1.7601]
版权所有 (c) 2009 Microsoft Corporation。保留所有权利。

C:\Users\Administrator>mysql -u root -p123456
mysql: [Warning] Using a password on the command line interface can be insecure.

Welcome to the MySQL monitor. Commands end with ; or \g.
Your MySQL connection id is 550
Server version: 5.7.20 MySQL Community Server (GPL)

Copyright (c) 2000, 2017, Oracle and/or its affiliates. All rights reserved.

Oracle is a registered trademark of Oracle Corporation and/or its
affiliates. Other names may be trademarks of their respective
owners.

Type 'help;' or '\h' for help. Type '\c' to clear the current input statement.

mysql> USE shop;
Database changed

mysql> SELECT * FROM user;
+----+------+-----+-----+----------+-------+-----------+--------+
| id | name | sex | age | password | phone | loginName | remark |
+----+------+-----+-----+----------+-------+-----------+--------+
3	david	女	28	111111	NULL	david	baann
4	小红	女	27	123456	NULL	xiaohong	black
5	小明	男	10	123456	NULL	xiaoming	berry
6	小刚	男	12	123456	NULL	xiaogang	banner
7	小王	男	14	111111	NULL	xiaowang	banana
8	小绿	女	34	222222	NULL	xiaolv	car
9	晓峰	男	15	333333	NULL	xiaofeng	carray
10	小影	女	26	444444	NULL	xiaoying	baaaa
11	大梅	女	27	555555	NULL	damei	accc
+----+------+-----+-----+----------+-------+-----------+--------+
9 rows in set (0.05 sec)

#按性别分组，并且查询在性别分组后大于2的性别，输出性别及数量
mysql> SELECT sex,count(sex) FROM user GROUP BY sex HAVING count(sex) >2;
+-----+------------+
| sex | count(sex) |
+-----+------------+
| 女 | 5 |
| 男 | 4 |
+-----+------------+
2 rows in set (0.07 sec)

#按性别分组，并且查询在性别分组后大于4的性别，输出性别及数量

```
mysql> SELECT sex,count(sex) FROM user GROUP BY sex HAVING count(sex) >4;
+-----+------------+
| sex | count(sex) |
+-----+------------+
| 女  |      5     |
+-----+------------+
1 row in set (0.00 sec)

#先用age>20的条件筛选，然后按性别分组，并且查询在性别分组后大于4的性别，输出性别及数量
mysql> SELECT sex,count(sex) FROM user WHERE age > 20 GROUP BY sex HAVING count(sex) >4;
+-----+------------+
| sex | count(sex) |
+-----+------------+
| 女  |      5     |
+-----+------------+
1 row in set (0.00 sec)

#先用age>27的条件筛选，然后按性别分组，并且查询在性别分组后大于4的性别，输出性别及数量
mysql> SELECT sex,count(sex) FROM user WHERE age > 27 GROUP BY sex HAVING count(sex) >4;
Empty set (0.00 sec)

mysql>
```

3.4.7 限制查询数量（LIMIT）

LIMIT用于限制查询的数量，常用于分页语句。LIMIT 子句可以被用于强制SELECT语句返回指定的记录数。LIMIT 接受一个或两个数字参数，参数必须是一个整数常量。

（1）如果只给定一个参数，则它表示返回最大的记录行数目。

```
#检索前6行记录
SELECT * FROM user LIMIT 6;
```

（2）如果给定两个参数，则第一个参数指定第一个返回记录行的偏移量，第二个参数指定返回记录行的最大数目，初始记录行的偏移量是0（而不是1）。

```
#从第2条数据开始，检索出5条数据
SELECT * FROM user LIMIT 2,5;
```

实战演练——数据检索

下面检索用户表前6条数据和从第3条数据开始的4条数据。

Microsoft Windows [版本 6.1.7601]
版权所有 (c) 2009 Microsoft Corporation。保留所有权利。

C:\Users\Administrator>mysql –u root –p123456
mysql: [Warning] Using a password on the command line interface can be insecure.

Welcome to the MySQL monitor. Commands end with ; or \g.
Your MySQL connection id is 551
Server version: 5.7.20 MySQL Community Server (GPL)

Copyright (c) 2000, 2017, Oracle and/or its affiliates. All rights reserved.

Oracle is a registered trademark of Oracle Corporation and/or its

affiliates. Other names may be trademarks of their respective
owners.

Type 'help;' or '\h' for help. Type '\c' to clear the current input statement.

mysql> USE shop;
Database changed

mysql> SELECT * FROM user;
+----+-------+------+------+----------+-------+-----------+--------+
| id | name | sex | age | password | phone | loginName | remark |
+----+-------+------+------+----------+-------+-----------+--------+
3	david	女	28	111111	NULL	david	baann
4	小红	女	27	123456	NULL	xiaohong	black
5	小明	男	10	123456	NULL	xiaoming	berry
6	小刚	男	12	123456	NULL	xiaogang	banner
7	小王	男	14	111111	NULL	xiaowang	banana
8	小绿	女	34	222222	NULL	xiaolv	car
9	晓峰	男	15	333333	NULL	xiaofeng	carray
10	小影	女	26	444444	NULL	xiaoying	baaaa
11	大梅	女	27	555555	NULL	damei	accc
+----+-------+------+------+----------+-------+-----------+--------+
9 rows in set (0.00 sec)

#检索前6条记录
mysql> SELECT * FROM user LIMIT 6;
+----+-------+------+------+----------+-------+-----------+--------+
| id | name | sex | age | password | phone | loginName | remark |
+----+-------+------+------+----------+-------+-----------+--------+
3	david	女	28	111111	NULL	david	baann
4	小红	女	27	123456	NULL	xiaohong	black
5	小明	男	10	123456	NULL	xiaoming	berry
6	小刚	男	12	123456	NULL	xiaogang	banner
7	小王	男	14	111111	NULL	xiaowang	banana
8	小绿	女	34	222222	NULL	xiaolv	car
+----+-------+------+------+----------+-------+-----------+--------+
6 rows in set (0.00 sec)

#检索从第3条记录开始的4条记录
mysql> SELECT * FROM user LIMIT 3,4;
+----+-------+------+------+----------+-------+-----------+--------+
| id | name | sex | age | password | phone | loginName | remark |
+----+-------+------+------+----------+-------+-----------+--------+
6	小刚	男	12	123456	NULL	xiaogang	banner
7	小王	男	14	111111	NULL	xiaowang	banana
8	小绿	女	34	222222	NULL	xiaolv	car
9	晓峰	男	15	333333	NULL	xiaofeng	carray
+----+-------+------+------+----------+-------+-----------+--------+
4 rows in set (0.00 sec)

#检索女生前3条数据
mysql> SELECT * FROM user WHERE sex='女' LIMIT 0,3;

```
+----+-------+-----+-----+----------+-------+-----------+--------+
| id | name  | sex | age | password | phone | loginName | remark |
+----+-------+-----+-----+----------+-------+-----------+--------+
|  3 | david | 女  |  28 | 111111   | NULL  | david     | baann  |
|  4 | 小红  | 女  |  27 | 123456   | NULL  | xiaohong  | black  |
|  8 | 小绿  | 女  |  34 | 222222   | NULL  | xiaolv    | car    |
+----+-------+-----+-----+----------+-------+-----------+--------+
3 rows in set (0.00 sec)

#检索女生按年龄排序后的前4条数据
mysql> SELECT * FROM user WHERE sex='女' ORDEY BY age DESC LIMIT 0,4;
+----+-------+-----+-----+----------+-------+-----------+--------+
| id | name  | sex | age | password | phone | loginName | remark |
+----+-------+-----+-----+----------+-------+-----------+--------+
|  8 | 小绿  | 女  |  34 | 222222   | NULL  | xiaolv    | car    |
|  3 | david | 女  |  28 | 111111   | NULL  | david     | baann  |
|  4 | 小红  | 女  |  27 | 123456   | NULL  | xiaohong  | black  |
| 11 | 大梅  | 女  |  27 | 555555   | NULL  | damei     | accc   |
+----+-------+-----+-----+----------+-------+-----------+--------+
4 rows in set (0.00 sec)

mysql>
```

3.5 字段操作

针对数据库表的字段，可以对字段进行设置为主键、设置为复合主键、添加字段、改变字段类型、字段重命名、字段设置默认值以及设置自增字段的操作。

精讲视频

字段操作

3.5.1 设置为主键

主键是一个表的唯一约束，是不可以为空、不可以重复的字段，如在外卖订单中，填写的手机号就可以作为主键。根据这个手机号，就可以找到相应的人，它起到唯一标识的作用。

创建表的时候，使用PRIMARY KEY添加主键。

CREATE TABLE tbl_name ([字段描述省略...], PRIMARY KEY(index_col_name));

实战演练——主键设置

创建一个学生表student，包含3个字段：id、name、no。先以id作为主键，然后再删除id主键，把no作为主键。主键也可以进行修改，前提是需要把前一个主键删除掉，才可以修改为新的主键。

```
Microsoft Windows [版本 6.1.7601]
版权所有 (c) 2009 Microsoft Corporation。保留所有权利。

C:\Users\Administrator>mysql -u root -p123456
mysql: [Warning] Using a password on the command line interface can be insecure.

Welcome to the MySQL monitor.  Commands end with ; or \g.
Your MySQL connection id is 52
Server version: 5.7.20 MySQL Community Server (GPL)

Copyright (c) 2000, 2017, Oracle and/or its affiliates. All rights reserved.
```

Oracle is a registered trademark of Oracle Corporation and/or its
affiliates. Other names may be trademarks of their respective
owners.

Type 'help;' or '\h' for help. Type '\c' to clear the current input statement.
mysql> USE shop;
Database changed

#创建表
mysql> CREATE TABLE student(
    -> id int not null,
    -> name varchar(255) not null,
    -> no int not null,
    -> primary key(id))
    -> ENGINE=Innodb DEFAULT CHARSET=utf8;
Query OK, 0 rows affected (0.60 sec)

#查看表主键
mysql> SHOW CREATE TABLE student;
+---------+---+
| Table | Create Table |
+---------+---+
| student | CREATE TABLE `student` (
  `id` int(11) NOT NULL,
  `name` varchar(255) NOT NULL,
  `no` int(11) NOT NULL,
  PRIMARY KEY (`id`)
) ENGINE=InnoDB DEFAULT CHARSET=utf8 |
+---------+---+
1 row in set (0.09 sec)

#删除表主键
mysql> ALTER TABLE student DROP primary key;
Query OK, 0 rows affected (0.96 sec)
Records: 0 Duplicates: 0 Warnings: 0

#表没有主键
mysql> SHOW CREATE TABLE student;
+---------+---+
| Table | Create Table |
+---------+---+
| student | CREATE TABLE `student` (
  `id` int(11) NOT NULL,
  `name` varchar(255) NOT NULL,
  `no` int(11) NOT NULL
) ENGINE=InnoDB DEFAULT CHARSET=utf8 |
+---------+---+
1 row in set (0.06 sec)

#设置no为主键
mysql> ALTER TABLE student add primary key(no);

```
Query OK, 0 rows affected (0.78 sec)
Records: 0  Duplicates: 0  Warnings: 0

mysql> SHOW CREATE TABLE student;
+---------+-----------------------------------------------------------+
| Table   | Create Table                                              |
+---------+-----------------------------------------------------------+
| student | CREATE TABLE `student` (
  `id` int(11) NOT NULL,
  `name` varchar(255) NOT NULL,
  `no` int(11) NOT NULL,
  PRIMARY KEY (`no`)
) ENGINE=InnoDB DEFAULT CHARSET=utf8 |
+---------+-----------------------------------------------------------+
1 row in set (0.00 sec)

mysql>
```

3.5.2 设置为复合主键

复合主键就是由多个字段组成的主键，它就像开启宝藏的钥匙，往往会分成两把或者更多把，当同时插入两把或更多把钥匙时，才能开启宝藏的大门，复合主键也是这样，以多个字段作为复合主键来确定唯一标识。

实战演练——复合主键设置

创建一个人员表person，包含3个字段：id、name、job，将id和name作为复合主键。

```
Microsoft Windows [版本 6.1.7601]
版权所有 (c) 2009 Microsoft Corporation。保留所有权利。

C:\Users\Administrator>mysql -u root -p123456
mysql: [Warning] Using a password on the command line interface can be insecure.

Welcome to the MySQL monitor.  Commands end with ; or \g.
Your MySQL connection id is 53
Server version: 5.7.20 MySQL Community Server (GPL)

Copyright (c) 2000, 2017, Oracle and/or its affiliates. All rights reserved.

Oracle is a registered trademark of Oracle Corporation and/or its
affiliates. Other names may be trademarks of their respective
owners.

Type 'help;' or '\h' for help. Type '\c' to clear the current input statement.

mysql> USE shop;
Database changed

mysql> CREATE TABLE person(
    -> id int not null,
    -> name varchar(255) not null,
    -> job varchar(255) not null,
    -> primary key(id,name))
```

```
    -> ENGINE=InnoDB DEFAULT CHARSET=utf8;
Query OK, 0 rows affected (0.63 sec)

mysql> SHOW CREATE TABLE person;
+--------+----------------------------------------------------------+
| Table  | Create Table                                             |
+--------+----------------------------------------------------------+
| person | CREATE TABLE `person` (
  `id` int(11) NOT NULL,
  `name` varchar(255) NOT NULL,
  `job` varchar(255) NOT NULL,
  PRIMARY KEY (`id`,`name`)
) ENGINE=InnoDB DEFAULT CHARSET=utf8 |
+--------+----------------------------------------------------------+
1 row in set (0.05 sec)

mysql>
```

3.5.3 添加字段

添加一个手机号码（phone）新字段到user表里，数据类型为字符串类型。

```
ALTER TABLE user add phone varchar(25) not Null;
```

实战演练——添加字段

```
Microsoft Windows [版本 6.1.7601]
版权所有 (c) 2009 Microsoft Corporation。保留所有权利。

C:\Users\Administrator>mysql –u root –p123456
mysql: [Warning] Using a password on the command line interface can be insecure.

Welcome to the MySQL monitor.  Commands end with ; or \g.
Your MySQL connection id is 39
Server version: 5.7.20 MySQL Community Server (GPL)

Copyright (c) 2000, 2017, Oracle and/or its affiliates. All rights reserved.

Oracle is a registered trademark of Oracle Corporation and/or its
affiliates. Other names may be trademarks of their respective
owners.

Type 'help;' or '\h' for help. Type '\c' to clear the current input statement.

mysql> USE shop;
Database changed
mysql> ALTER TABLE user add phone varchar(25) not Null;
Query OK, 0 rows affected (0.98 sec)
Records: 0  Duplicates: 0  Warnings: 0

mysql> DESC user;
+-------+------+------+-----+---------+-------+
| Field | Type | Null | Key | Default | Extra |
```

```
+----------+------------------+------+-----+---------+----------------+
| id       | int(10) unsigned | NO   | PRI | NULL    | auto_increment |
| name     | varchar(25)      | YES  |     | NULL    |                |
| sex      | varchar(5)       | YES  |     | NULL    |                |
| age      | int(10)          | YES  |     | NULL    |                |
| password | varchar(25)      | YES  |     | NULL    |                |
| phone    | varchar(25)      | NO   |     | NULL    |                |
+----------+------------------+------+-----+---------+----------------+
6 rows in set (0.00 sec)

mysql>
```

3.5.4 改变字段类型

可以修改表字段的数据类型，将手机号码（phone）字符串类型修改为整型（int）。

ALTER TABLE user modify phone int(25) not Null;

实战演练——改变字段类型

```
Microsoft Windows [版本 6.1.7601]
版权所有 (c) 2009 Microsoft Corporation。保留所有权利。

C:\Users\Administrator>mysql –u root –p123456
mysql: [Warning] Using a password on the command line interface can be insecure.

Welcome to the MySQL monitor.  Commands end with ; or \g.
Your MySQL connection id is 40
Server version: 5.7.20 MySQL Community Server (GPL)

Copyright (c) 2000, 2017, Oracle and/or its affiliates. All rights reserved.

Oracle is a registered trademark of Oracle Corporation and/or its
affiliates. Other names may be trademarks of their respective
owners.

Type 'help;' or '\h' for help. Type '\c' to clear the current input statement.

mysql> USE shop;
Database changed
mysql> ALTER TABLE user modify phone int(25) not Null;
Query OK, 0 rows affected (1.59 sec)
Records: 0  Duplicates: 0  Warnings: 0

mysql> DESC user;
+----------+------------------+------+-----+---------+----------------+
| Field    | Type             | Null | Key | Default | Extra          |
+----------+------------------+------+-----+---------+----------------+
| id       | int(10) unsigned | NO   | PRI | NULL    | auto_increment |
| name     | varchar(25)      | YES  |     | NULL    |                |
| sex      | varchar(5)       | YES  |     | NULL    |                |
| age      | int(10)          | YES  |     | NULL    |                |
| password | varchar(25)      | YES  |     | NULL    |                |
```

```
| phone    | int(25)       | NO  |     | NULL    |       |
+----------+---------------+-----+-----+---------+-------+
6 rows in set (0.00 sec)

mysql>
```

3.5.5 字段重命名

对于已经存在的表结构，如果想对表里的字段重命名，则需要使用alter table来修改表里的字段，格式如下所示。

ALTER TABLE <表名> change <字段名> <字段新名称> <字段的类型>

实战演练 ——字段重命名

下面将user 表里的phone字段改为telephone字段。

Microsoft Windows [版本 6.1.7601]
版权所有 (c) 2009 Microsoft Corporation。保留所有权利。

C:\Users\Administrator>mysql –u root –p123456
mysql: [Warning] Using a password on the command line interface can be insecure.

Welcome to the MySQL monitor. Commands end with ; or \g.
Your MySQL connection id is 54
Server version: 5.7.20 MySQL Community Server (GPL)

Copyright (c) 2000, 2017, Oracle and/or its affiliates. All rights reserved.

Oracle is a registered trademark of Oracle Corporation and/or its
affiliates. Other names may be trademarks of their respective
owners.

Type 'help;' or '\h' for help. Type '\c' to clear the current input statement.

mysql> USE shop;
Database changed

mysql> DESC user;
```
+----------+-----------------+-----+-----+---------+----------------+
| Field    | Type            | Null| Key | Default | Extra          |
+----------+-----------------+-----+-----+---------+----------------+
| id       | int(10) unsigned| NO  | PRI | NULL    | auto_increment |
| name     | varchar(25)     | YES |     | NULL    |                |
| sex      | varchar(5)      | YES |     | NULL    |                |
| age      | int(10)         | YES |     | NULL    |                |
| password | varchar(25)     | YES |     | NULL    |                |
| phone    | int(25)         | NO  |     | NULL    |                |
+----------+-----------------+-----+-----+---------+----------------+
6 rows in set (0.02 sec)
```

mysql> ALTER TABLE user change phone telephone int(25);
Query OK, 0 rows affected (0.96 sec)

Records: 0 Duplicates: 0 Warnings: 0

```
mysql> DESC user;
+-----------+--------------+------+-----+---------+----------------+
| Field     | Type         | Null | Key | Default | Extra          |
+-----------+--------------+------+-----+---------+----------------+
| id        | int(10) unsigned | NO  | PRI | NULL   | auto_increment |
| name      | varchar(25)  | YES  |     | NULL    |                |
| sex       | varchar(5)   | YES  |     | NULL    |                |
| age       | int(10)      | YES  |     | NULL    |                |
| password  | varchar(25)  | YES  |     | NULL    |                |
| telephone | int(25)      | YES  |     | NULL    |                |
+-----------+--------------+------+-----+---------+----------------+
6 rows in set (0.00 sec)

mysql>
```

3.5.6 字段设置默认值

MySQL数据库字段在创建的时候可以设置默认值，也可以修改它的默认值，如果有默认值，也可以将默认值删除。

设置默认值的命令如下所示。

ALTER TABLE 表名 ALTER 字段名 SET default 默认值；

删除默认值的命令如下所示。

ALTER TABLE 表名 ALTER 字段名 DROP default;

实战演练 ——字段设置默认值

创建一个部门表dept，包含3个字段：id、deptName、userName，部门名称默认值为软件事业部。

Microsoft Windows [版本 6.1.7601]
版权所有 (c) 2009 Microsoft Corporation。保留所有权利。

C:\Users\Administrator>mysql –u root –p123456
mysql: [Warning] Using a password on the command line interface can be insecure.

Welcome to the MySQL monitor. Commands end with ; or \g.
Your MySQL connection id is 57
Server version: 5.7.20 MySQL Community Server (GPL)

Copyright (c) 2000, 2017, Oracle and/or its affiliates. All rights reserved.

Oracle is a registered trademark of Oracle Corporation and/or its
affiliates. Other names may be trademarks of their respective
owners.

Type 'help;' or '\h' for help. Type '\c' to clear the current input statement.

#进入shop数据库
mysql> USE shop;
Database changed

#创建dept部门表

```
mysql> CREATE TABLE dept(
    -> id int not null,
    -> deptName varchar(255) not null default '软件事业部',
    -> userName varchar(25) not null,
    -> primary key(id))
    -> ENGINE=InnoDB default charset=utf8;
Query OK, 0 rows affected (0.41 sec)

#插入数据,但不插入部门名称数据
mysql> INSERT INTO dept(id,userName) VALUES(1,'kevin');
Query OK, 1 row affected (0.02 sec)

#查询dept表,看到没有插入deptName部门名称的值,但是它会插入默认值
mysql> SELECT * FROM dept;
+----+--------------+----------+
| id | deptName     | userName |
+----+--------------+----------+
|  1 | 软件事业部   | kevin    |
+----+--------------+----------+
1 row in set (0.00 sec)

#修改dept表的默认值
mysql> ALTER TABLE dept ALTER deptName SET default '办公室';
Query OK, 0 rows affected (0.10 sec)
Records: 0  Duplicates: 0  Warnings: 0

#插入数据,但不插入部门名称数据
mysql> INSERT INTO dept(id,userName) VALUES(2,'tom');
Query OK, 1 row affected (0.00 sec)

#查询dept表,看到没有插入deptName部门名称的值,但是它会插入新的默认值
mysql> SELECT * FROM dept;
+----+--------------+----------+
| id | deptName     | userName |
+----+--------------+----------+
|  1 | 软件事业部   | kevin    |
|  2 | 办公室       | tom      |
+----+--------------+----------+
2 rows in set (0.01 sec)

#删除deptName部门名称的默认值
mysql> ALTER TABLE dept ALTER deptName DROP default;
Query OK, 0 rows affected (0.12 sec)
Records: 0  Duplicates: 0  Warnings: 0

#插入数据时,如果没有默认值,则不让插入
mysql> INSERT INTO dept(id,userName) VALUES(3,'david');
ERROR 1364 (HY000): Field 'deptName' doesn't have a default value

#插入数据
mysql> INSERT INTO dept(id,deptName,userName) VALUES(3,'人力资源部','david');
Query OK, 1 row affected (0.00 sec)
```

```
#查询数据
mysql> SELECT * FROM dept;
+----+--------------+----------+
| id | deptName     | username |
+----+--------------+----------+
|  1 | 软件事业部   | kevin    |
|  2 | 办公室       | tom      |
|  3 | 人力资源部   | david    |
+----+--------------+----------+
3 rows in set (0.00 sec)

mysql>
```

3.5.7　设置自增字段

　　MySQL数据库表经常会将主键设置为自增字段。从设计的角度来说，表的主键应尽量设计成一个与业务无关的字段，如果将它设计成自增，则应用系统在开发的时候不用关心这个主键的设定，由数据库自己来维护，但是有在高并发下成为瓶颈的风险，当然如果并发并不是非常高的话，一般不会成为瓶颈。如果自己控制这个主键值，则更需要付出一点代价来生成这个值，并发问题可以通过扩展应用集群来解决。

　　在MySQL中定义字段列为自增的属性：AUTO_INCREMENT。

　　（1）如果把一个NULL插入到一个AUTO_INCREMENT数据列里去，MySQL将自动生成下一个序列编号，编号从1开始，并以1为基数递增。

　　（2）当插入记录时，没有为AUTO_INCREMENT明确指定值，则等同于插入NULL值。

　　（3）当插入记录时，如果为AUTO_INCREMENT字段明确指定了一个数值，则会出现两种情况：一种是如果插入的值与已有的编号重复，则会出现出错信息，因为AUTO_INCREMENT数据列的值必须是唯一的；另一种是如果插入的值大于已编号的值，则会把该值插入到数据列中，下一个编号将从这个新值开始递增。

　　（4）对于使用MyISAM存储引擎，如果用UPDATE命令更新自增列，若列值与已有值重复，则会出错，若列值大于已有值，则下一个编号从该值开始递增；但是对于Innodb存储引擎，UPDATE auto_increment字段会导致发生报错。

　　（5）被DELETE语句删除的id值，除非在SQL中将id重新插入，否则前面空余的id不会复用。

实战演练——设置自增字段

Microsoft Windows [版本 6.1.7601]
版权所有 (c) 2009 Microsoft Corporation。保留所有权利。
#输入用户名、密码，进入MySQL数据库
C:\Users\Administrator>mysql –u root –p123456
mysql: [Warning] Using a password on the command line interface can be insecure.

Welcome to the MySQL monitor. Commands end with ; or \g.
Your MySQL connection id is 10
Server version: 5.7.20 MySQL Community Server (GPL)

Copyright (c) 2000, 2017, Oracle and/or its affiliates. All rights reserved.

Oracle is a registered trademark of Oracle Corporation and/or its
affiliates. Other names may be trademarks of their respective
owners.

```
Type 'help;' or '\h' for help. Type '\c' to clear the current input statement.
#使用shop数据库
mysql> USE shop;
Database changed

#创建t_zizeng表，设置id为自增字段，但是没有设置成主键，会报错
mysql> CREATE TABLE t_zizeng(id int auto_increment,name varchar(255));
ERROR 1075 (42000): Incorrect table definition; there can be only one auto colum
n and it must be defined as a key

#创建t_zizeng表，将id设置成自增字段，并且把它设置为主键
mysql> CREATE TABLE t_zizeng(id int auto_increment,name varchar(255),primary key
(id));
Query OK, 0 rows affected (0.32 sec)

#插入一条数据，自增字段的值为null，它会自动生成
mysql> INSERT INTO t_zizeng(id,name) VALUES(null,'小明');
Query OK, 1 row affected (0.00 sec)

#查询t_zizeng表，可以看到id值自动生成为1
mysql> SELECT * FROM t_zizeng;
+----+--------+
| id | name   |
+----+--------+
|  1 | 小明   |
+----+--------+
1 row in set (0.00 sec)

#插入一条数据，自增字段可以不用赋值，它会自动增加
mysql> INSERT INTO t_zizeng(name) VALUES('小刚');
Query OK, 1 row affected (0.00 sec)

#查询t_zizeng表，可以看到id值自动生成为2
mysql> SELECT * FROM t_zizeng;
+----+--------+
| id | name   |
+----+--------+
|  1 | 小明   |
|  2 | 小刚   |
+----+--------+
2 rows in set (0.00 sec)

#插入重复的主键id值，它会报错
mysql> INSERT INTO t_zizeng(id,name) VALUES(1,'小王');
ERROR 1062 (23000): Duplicate entry '1' for key 'PRIMARY'

#手动插入主键id值100，下次自动生成id值时会以100为基数
mysql> INSERT INTO t_zizeng(id,name) VALUES(100,'小王');
Query OK, 1 row affected (0.00 sec)

#插入一条数据，自增字段可以不用赋值，它会自动增加
```

```
mysql> INSERT INTO t_zizeng(name) VALUES('小红');
Query OK, 1 row affected (0.00 sec)

#查询t_zizeng表,可以看到id值自动生成为101
mysql> SELECT * FROM t_zizeng;
+-----+--------+
| id  | name   |
+-----+--------+
|   1 | 小明   |
|   2 | 小刚   |
| 100 | 小王   |
| 101 | 小红   |
+-----+--------+
4 rows in set (0.00 sec)

#用DELETE语句删除的id值,除非在SQL中将id重新插入,否则前面空余的id不会复用
mysql> DELETE FROM t_zizeng WHERE id=101;
Query OK, 1 row affected (0.05 sec)

#插入一条数据,自增字段可以不用赋值,它会自动增加
mysql> INSERT INTO t_zizeng(name) VALUES('小张');
Query OK, 1 row affected (0.00 sec)

#可以看到删除的101没有再次使用,而是使用102
mysql> SELECT * FROM t_zizeng;
+-----+--------+
| id  | name   |
+-----+--------+
|   1 | 小明   |
|   2 | 小刚   |
| 100 | 小王   |
| 102 | 小张   |
+-----+--------+
4 rows in set (0.00 sec)

mysql>
```

3.6 客户端操作数据库

MySQL数据库不光可以使用命令行来进行数据库的创建、数据库的删除、表的创建、表的删除以及数据的插入、删除、修改、查询等操作,同时也可以使用界面的客户端来操作,下面就使用客户端来操作MySQL数据库。

实战演练——使用客户端

(1) 下载MySQL客户端,然后进行安装。
(2) 输入用户名、密码,连接本地MySQL数据库,如图3.3所示。
(3) 在MySQL客户端界面中,经常会用到左侧的数据库区域、输入命令窗口区域、输出结果区域,如图3.4所示。

精讲视频

客户端操作
数据库

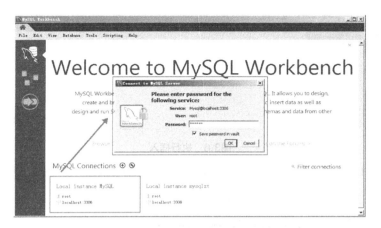

图3.3　连接本地MySQL数据库

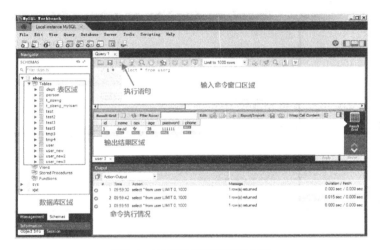

图3.4　MySQL客户端界面介绍

（4）在MySQL客户端中插入一条数据，可以单击"新增数据"按钮，编写要插入的数据，最后单击"执行"按钮，如图3.5和图3.6所示。

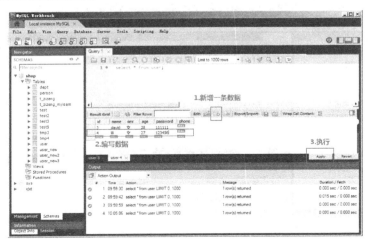

图3.5　插入数据

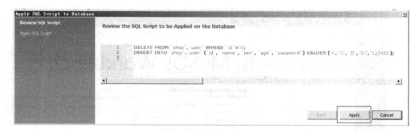

图3.6 执行插入

（5）在MySQL客户端中删除一条数据，需要选中要删除的行数据，然后按"删除"按钮，如图3.7所示。

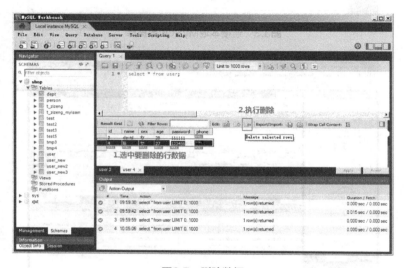

图3.7 删除数据

（6）在MySQL客户端中修改数据，需要选中要修改的行数据，然后单击"修改"按钮，编辑行数据，最后按"执行"按钮，如图3.8所示。

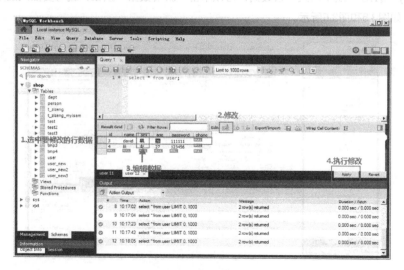

图3.8 修改数据

（7）在MySQL客户端中，通过命令窗口区域，可以输入MySQL命令进行操作，包括修改数据、插入数据、删除数据、查询数据等，如图3.9所示。如果要注释命令，可以使用空格键。

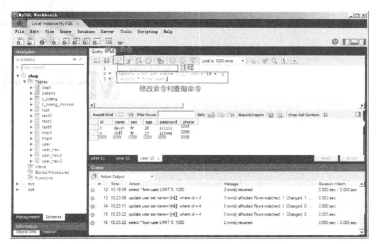

图3.9　命令操作

（8）在MySQL客户端中，可以创建数据库以及数据库表，如图3.10所示。

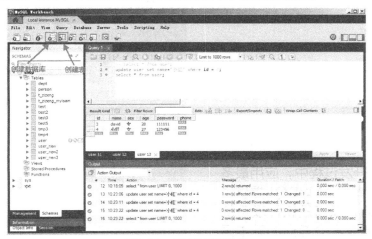

图3.10　创建数据库及表

除了以上介绍的操作外，MySQL客户端还可以进行很多其他操作，如保存MySQL命令等，大家可以自己尝试。

3.7　小结

本章主要讲述了MySQL的常用操作，这是必须要掌握的内容。通过本章，我们学会了数据库用户的管理，包括连接MySQL数据库、新增用户、修改用户密码的操作；学会了数据库的操作，包括创建数据库、查看数据库、使用数据库以及删除数据库；学会了表的操作，包括创建表、查看表结构、复制表、临时表以及内存表的操作；学会了数据的常用操作，包括插入数据、查询数据、修改数据、删除数据、对查询结果排序、对查询结果分组、设置分组条件、限制查询数量；学会了字段的常用操作，包括设置为主键、设置为复合主键、添加字段、改变字段类型、字段重命名、字段设置默认值、自增字段的使用；最后学会了使用MySQL客户端来进行数据库操作。

第4章
MySQL查询

本章要点

- 基本查询语法
- 数据过滤
- 子查询
- 聚合函数
- 高级查询
- 小结

■ 在软件项目开发过程中，应用最多的就是MySQL查询，MySQL提供了各种场景的查询，以满足软件项目的开发要求，本章要学会查询的基本语法、数据过滤、子查询、聚合分组查询、连接查询以及组合查询，这些也是MySQL必须要掌握的内容，熟练使用这些查询，在项目开发过程中就会得心应手，提高开发效率。

4.1 基本查询语法

MySQL使用关键字SELECT来进行查询，SELECT语句的基本语法格式如下。

SELECT 查询内容
FROM 表名
WHERE 表达式
GROUP BY 字段名
HAVING 表达式
ORDEY BY 字段名
LIMIT 记录数

精讲视频

基本查询语法

（1）SELECT 查询内容：查询所有字段 SELECT * FROM 表名，*通配符表示所有字段。

（2）FROM 表名：指定是从哪张表中查询。

（3）WHERE 表达式（按条件查询）：在MySQL的表查询时，往往并不是需要将所有内容全部查出，而是根据实际需求，查询所需数据。

在MySQL语句中，条件表达式是指SELECT语句的查询条件，在WHERE子句中可以使用关系运算符连接操作数作为查询条件对数据进行选择。常见的关系运算符如下。

= 等于

<> 不等于

!= 不等于

< 小于

> 大于

<= 小于等于

>= 大于等于

（4）GROUP BY字段名：分组进行查询，GROUP BY 把符合条件的同一类分在一组，一般GROUP BY是和聚合函数配合使用的，如按男生分组查询、按女生分组查询。

（5）HAVING 表达式：HAVING关键字都用于设置条件表达式对查询结果进行过滤，区别是HAVING后面可以跟聚合函数，而WHERE不能，通常HAVING关键字都与GROUP BY一起使用，表示对分组后的数据进行过滤。

（6）ORDEY BY字段名：按某个字段进行排序，如淘宝宝贝可以按销量排序、按好评数排序、按价格排序等。

（7）LIMIT记录数：查询的条数，可以限制每次查询多少条数据，进行分页查询。

实战演练 —— 查询数据

Microsoft Windows [版本 6.1.7601]
版权所有 (c) 2009 Microsoft Corporation。保留所有权利。

#输入用户名、密码，进入数据库
C:\Users\Administrator>mysql –u root –p123456
mysql: [Warning] Using a password on the command line interface can be insecure.

Welcome to the MySQL monitor. Commands end with ; or \g.
Your MySQL connection id is 20

```
Server version: 5.7.20 MySQL Community Server (GPL)

Copyright (c) 2000, 2017, Oracle and/or its affiliates. All rights reserved.

Oracle is a registered trademark of Oracle Corporation and/or its
affiliates. Other names may be trademarks of their respective
owners.

Type 'help;' or '\h' for help. Type '\c' to clear the current input statement.

#使用shop数据库
mysql> USE shop;
Database changed

#查询user表里的数据，显示所有字段的数据
mysql> SELECT * FROM user;
+----+------+-----+-----+----------+-------+
| id | name | sex | age | password | phone |
+----+------+-----+-----+----------+-------+
|  3 | david| 女  |  28 | 111111   | NULL  |
|  4 | 小红 | 女  |  27 | 123456   | NULL  |
|  5 | 小明 | 男  |  10 | 123456   | NULL  |
|  6 | 小刚 | 男  |  12 | 123456   | NULL  |
|  7 | 小王 | 男  |  14 | 111111   | NULL  |
|  8 | 小绿 | 女  |  34 | 222222   | NULL  |
|  9 | 晓峰 | 男  |  15 | 333333   | NULL  |
| 10 | 小影 | 女  |  26 | 444444   | NULL  |
| 11 | 大梅 | 女  |  27 | 555555   | NULL  |
+----+------+-----+-----+----------+-------+
9 rows in set (0.00 sec)

#查询user表里的数据，只显示name字段的数据
mysql> SELECT name FROM user;
+------+
| name |
+------+
| david|
| 小红 |
| 小明 |
| 小刚 |
| 小王 |
| 小绿 |
| 晓峰 |
| 小影 |
| 大梅 |
+------+
9 rows in set (0.00 sec)

#按条件进行查询，查询性别等于男的数据
mysql> SELECT * FROM user WHERE sex='男';
+----+------+-----+-----+----------+-------+
| id | name | sex | age | password | phone |
```

```
+----+--------+-----+-----+----------+------+
| 5  | 小明   | 男  | 10  | 123456   | NULL |
| 6  | 小刚   | 男  | 12  | 123456   | NULL |
| 7  | 小王   | 男  | 14  | 111111   | NULL |
| 9  | 晓峰   | 男  | 15  | 333333   | NULL |
+----+--------+-----+-----+----------+------+
4 rows in set (0.00 sec)
```

#查询年龄大于20的数据
```
mysql> SELECT * FROM user WHERE age > 20;
+----+--------+-----+-----+----------+------+
| id | name   | sex | age | password | phone|
+----+--------+-----+-----+----------+------+
| 3  | david  | 女  | 28  | 111111   | NULL |
| 4  | 小红   | 女  | 27  | 123456   | NULL |
| 8  | 小绿   | 女  | 34  | 222222   | NULL |
| 10 | 小影   | 女  | 26  | 444444   | NULL |
| 11 | 大梅   | 女  | 27  | 555555   | NULL |
+----+--------+-----+-----+----------+------+
5 rows in set (0.02 sec)
```

#查询总共有多少条数据
```
mysql> SELECT count(*) FROM user;
+----------+
| count(*) |
+----------+
|    9     |
+----------+
1 row in set (0.08 sec)
```

#查询年龄总和
```
mysql> SELECT sum(age) FROM user;
+----------+
| sum(age) |
+----------+
|   193    |
+----------+
1 row in set (0.05 sec)
```

#查询年龄最大的数据
```
mysql> SELECT max(age) FROM user;
+----------+
| max(age) |
+----------+
|    34    |
+----------+
1 row in set (0.04 sec)
```

#按性别进行分组查询
```
mysql> SELECT * FROM user GROUP BY sex;
+----+--------+-----+-----+----------+------+
| id | name   | sex | age | password | phone|
```

```
+--+----+----+----+------+----+
| 3 | david | 女 | 28 | 111111 | NULL |
| 5 | 小明  | 男 | 10 | 123456 | NULL |
+--+----+----+----+------+----+
```
2 rows in set (0.06 sec)

#按年龄降序排列查询
mysql> SELECT * FROM user ORDEY BY age DESC;
```
+--+----+----+----+------+----+
| id | name | sex | age | password | phone |
+--+----+----+----+------+----+
| 8  | 小绿  | 女 | 34 | 222222 | NULL |
| 3  | david | 女 | 28 | 111111 | NULL |
| 4  | 小红  | 女 | 27 | 123456 | NULL |
| 11 | 大梅  | 女 | 27 | 555555 | NULL |
| 10 | 小影  | 女 | 26 | 444444 | NULL |
| 9  | 晓峰  | 男 | 15 | 333333 | NULL |
| 7  | 小王  | 男 | 14 | 111111 | NULL |
| 6  | 小刚  | 男 | 12 | 123456 | NULL |
| 5  | 小明  | 男 | 10 | 123456 | NULL |
+--+----+----+----+------+----+
```
9 rows in set (0.00 sec)

#按年龄升序排列查询
mysql> SELECT * FROM user ORDEY BY age ASC;
```
+--+----+----+----+------+----+
| id | name | sex | age | password | phone |
+--+----+----+----+------+----+
| 5  | 小明  | 男 | 10 | 123456 | NULL |
| 6  | 小刚  | 男 | 12 | 123456 | NULL |
| 7  | 小王  | 男 | 14 | 111111 | NULL |
| 9  | 晓峰  | 男 | 15 | 333333 | NULL |
| 10 | 小影  | 女 | 26 | 444444 | NULL |
| 4  | 小红  | 女 | 27 | 123456 | NULL |
| 11 | 大梅  | 女 | 27 | 555555 | NULL |
| 3  | david | 女 | 28 | 111111 | NULL |
| 8  | 小绿  | 女 | 34 | 222222 | NULL |
+--+----+----+----+------+----+
```
9 rows in set (0.00 sec)

#查询前5条数据
mysql> SELECT * FROM user LIMIT 0,5;
```
+--+----+----+----+------+----+
| id | name | sex | age | password | phone |
+--+----+----+----+------+----+
| 3 | david | 女 | 28 | 111111 | NULL |
| 4 | 小红  | 女 | 27 | 123456 | NULL |
| 5 | 小明  | 男 | 10 | 123456 | NULL |
| 6 | 小刚  | 男 | 12 | 123456 | NULL |
| 7 | 小王  | 男 | 14 | 111111 | NULL |
+--+----+----+----+------+----+
```
5 rows in set (0.00 sec)

mysql>

4.2 数据过滤

数据过滤用在WHERE表达式里，常用的有基本查询过滤、条件查询过滤、模糊查询过滤、字段控制查询过滤以及正则表达式查询过滤。

4.2.1 基本查询过滤

基本查询过滤可以用来查询所有字段数据或指定一个字段或者多个字段的数据。

实战演练——基本查询过滤

Microsoft Windows [版本 6.1.7601]
版权所有 (c) 2009 Microsoft Corporation。保留所有权利。

C:\Users\Administrator>mysql –u root –p123456
mysql: [Warning] Using a password on the command line interface can be insecure.

Welcome to the MySQL monitor. Commands end with ; or \g.
Your MySQL connection id is 21
Server version: 5.7.20 MySQL Community Server (GPL)

Copyright (c) 2000, 2017, Oracle and/or its affiliates. All rights reserved.

Oracle is a registered trademark of Oracle Corporation and/or its
affiliates. Other names may be trademarks of their respective
owners.

Type 'help;' or '\h' for help. Type '\c' to clear the current input statement.

mysql> USE shop;
Database changed

#查询所有字段数据
mysql> SELECT * FROM user;

id	name	sex	age	password	phone
3	david	女	28	111111	NULL
4	小红	女	27	123456	NULL
5	小明	男	10	123456	NULL
6	小刚	男	12	123456	NULL
7	小王	男	14	111111	NULL
8	小绿	女	34	222222	NULL
9	晓峰	男	15	333333	NULL
10	小影	女	26	444444	NULL
11	大梅	女	27	555555	NULL

9 rows in set (0.00 sec)

```
#查询指定一个或多个字段数据
mysql> SELECT name,sex,age FROM user;
+------+-----+-----+
| name | sex | age |
+------+-----+-----+
| david| 女  | 28  |
| 小红 | 女  | 27  |
| 小明 | 男  | 10  |
| 小刚 | 男  | 12  |
| 小王 | 男  | 14  |
| 小绿 | 女  | 34  |
| 晓峰 | 男  | 15  |
| 小影 | 女  | 26  |
| 大梅 | 女  | 27  |
+------+-----+-----+
9 rows in set (0.00 sec)
```

4.2.2 条件查询过滤

条件查询过滤关键字为AND、OR、IN、NOT IN、IS NULL、IS NOT NULL、BETWEEN AND。按关键字可以进行条件查询过滤。

（1）AND（与）。用AND进行查询的时候，查询出来的数据要求条件都得满足。

SELECT * FROM user WHERE age = 20 and name='kevin';

（2）OR（或）。用OR进行查询的时候，查询出来的数据只要求满足任意一个条件就可以查询出来。

SELECT * FROM user WHERE age = 20 or name='kevin';

（3）IN（在范围内）。用IN进行查询的时候，查询出来的数据在这个范围内。

SELECT * FROM user WHERE id in (3,5,7);

（4）NOT IN（不在范围内）。用NOT IN进行查询的时候，查询出来的数据不在这个范围内。

SELECT * FROM user WHERE id not in(3,5,7);

（5）IS（为空）。用IS NULL进行查询的时候，用来查询某字段为空时用is null，而不能使用"=null"，因为MySQL中的null不等于任何其他值，也不等于另外一个null，优化器会把"=null"的查询过滤掉而不返回任何数据；查询某字段为非空时使用is not null。

SELECT * FROM user WHERE name is null;

（6）BETWEEN AND（在……区间）：用BETWEEN AND进行查询的时候，查询出来的数据在这个区间。

SELECT * FROM user WHERE age between 10 and 20;

实战演练——条件查询过滤

Microsoft Windows [版本 6.1.7601]
版权所有 (c) 2009 Microsoft Corporation。保留所有权利。

C:\Users\Administrator>mysql -u root -p123456
mysql: [Warning] Using a password on the command line interface can be insecure.

Welcome to the MySQL monitor. Commands end with ; or \g.
Your MySQL connection id is 24
Server version: 5.7.20 MySQL Community Server (GPL)

Copyright (c) 2000, 2017, Oracle and/or its affiliates. All rights reserved.

Oracle is a registered trademark of Oracle Corporation and/or its affiliates. Other names may be trademarks of their respective owners.

Type 'help;' or '\h' for help. Type '\c' to clear the current input statement.

mysql> USE shop;
Database changed

#无条件查询所有数据
mysql> SELECT * FROM user;

id	name	sex	age	password	phone
3	david	女	28	111111	NULL
4	小红	女	27	123456	NULL
5	小明	男	10	123456	NULL
6	小刚	男	12	123456	NULL
7	小王	男	14	111111	NULL
8	小绿	女	34	222222	NULL
9	晓峰	男	15	333333	NULL
10	小影	女	26	444444	NULL
11	大梅	女	27	555555	NULL

9 rows in set (0.00 sec)

#使用AND，查询性别为女并且年龄大于27岁的数据
mysql> SELECT * FROM user WHERE sex='女' and age > 27;

id	name	sex	age	password	phone
3	david	女	28	111111	NULL
8	小绿	女	34	222222	NULL

2 rows in set (0.00 sec)

#使用OR，查询姓名等于小影或者姓名等于小明的数据
mysql> SELECT * FROM user WHERE name ='小影' or name='小明';

id	name	sex	age	password	phone
5	小明	男	10	123456	NULL
10	小影	女	26	444444	NULL

2 rows in set (0.00 sec)

#使用IN，查询id在（3,5,7）范围内的数据
mysql> SELECT * FROM user WHERE id in(3,5,7);

id	name	sex	age	password	phone

```
| 3 | david | 女 | 28 | 111111 | NULL |
| 5 | 小明  | 男 | 10 | 123456 | NULL |
| 7 | 小王  | 男 | 14 | 111111 | NULL |
+----+-------+-----+-----+----------+-------+
3 rows in set (0.06 sec)
```

#使用NOT IN，查询id不在（3,5,7）范围内的数据
```
mysql> SELECT * FROM user WHERE id not in(3,5,7);
+----+-------+-----+-----+----------+-------+
| id | name  | sex | age | password | phone |
+----+-------+-----+-----+----------+-------+
| 4  | 小红  | 女  | 27  | 123456   | NULL  |
| 6  | 小刚  | 男  | 12  | 123456   | NULL  |
| 8  | 小绿  | 女  | 34  | 222222   | NULL  |
| 9  | 晓峰  | 男  | 15  | 333333   | NULL  |
| 10 | 小影  | 女  | 26  | 444444   | NULL  |
| 11 | 大梅  | 女  | 27  | 555555   | NULL  |
+----+-------+-----+-----+----------+-------+
6 rows in set (0.00 sec)
```

#使用IS NULL，查询手机号码为空的数据
```
mysql> SELECT * FROM user WHERE phone is null;
+----+-------+-----+-----+----------+-------+
| id | name  | sex | age | password | phone |
+----+-------+-----+-----+----------+-------+
| 3  | david | 女  | 28  | 111111   | NULL  |
| 4  | 小红  | 女  | 27  | 123456   | NULL  |
| 5  | 小明  | 男  | 10  | 123456   | NULL  |
| 6  | 小刚  | 男  | 12  | 123456   | NULL  |
| 7  | 小王  | 男  | 14  | 111111   | NULL  |
| 8  | 小绿  | 女  | 34  | 222222   | NULL  |
| 9  | 晓峰  | 男  | 15  | 333333   | NULL  |
| 10 | 小影  | 女  | 26  | 444444   | NULL  |
| 11 | 大梅  | 女  | 27  | 555555   | NULL  |
+----+-------+-----+-----+----------+-------+
9 rows in set (0.22 sec)
```

#使用IS NOT NULL，查询手机号码不为空的数据
```
mysql> SELECT * FROM user WHERE phone is not null;
Empty set (0.00 sec)
```

#查询年龄大于20并且小于30的数据
```
mysql> SELECT * FROM user WHERE age > 20 and age < 30;
+----+-------+-----+-----+----------+-------+
| id | name  | sex | age | password | phone |
+----+-------+-----+-----+----------+-------+
| 3  | david | 女  | 28  | 111111   | NULL  |
| 4  | 小红  | 女  | 27  | 123456   | NULL  |
| 10 | 小影  | 女  | 26  | 444444   | NULL  |
| 11 | 大梅  | 女  | 27  | 555555   | NULL  |
+----+-------+-----+-----+----------+-------+
4 rows in set (0.00 sec)
```

#使用BTWEEN AND，查询年龄大于20并且小于30的数据
```
mysql> SELECT * FROM user WHERE age between 20 and 30;
+----+-------+-----+------+----------+-------+
| id | name  | sex | age  | password | phone |
+----+-------+-----+------+----------+-------+
|  3 | david | 女  |   28 | 111111   | NULL  |
|  4 | 小红  | 女  |   27 | 123456   | NULL  |
| 10 | 小影  | 女  |   26 | 444444   | NULL  |
| 11 | 大梅  | 女  |   27 | 555555   | NULL  |
+----+-------+-----+------+----------+-------+
4 rows in set (0.06 sec)
```

#查询性别为男的数据
```
mysql> SELECT * FROM user WHERE sex = '男';
+----+-------+-----+------+----------+-------+
| id | name  | sex | age  | password | phone |
+----+-------+-----+------+----------+-------+
|  5 | 小明  | 男  |   10 | 123456   | NULL  |
|  6 | 小刚  | 男  |   12 | 123456   | NULL  |
|  7 | 小王  | 男  |   14 | 111111   | NULL  |
|  9 | 晓峰  | 男  |   15 | 333333   | NULL  |
+----+-------+-----+------+----------+-------+
4 rows in set (0.00 sec)
```

#查询性别不为男的数据
```
mysql> SELECT * FROM user WHERE sex != '男';
+----+-------+-----+------+----------+-------+
| id | name  | sex | age  | password | phone |
+----+-------+-----+------+----------+-------+
|  3 | david | 女  |   28 | 111111   | NULL  |
|  4 | 小红  | 女  |   27 | 123456   | NULL  |
|  8 | 小绿  | 女  |   34 | 222222   | NULL  |
| 10 | 小影  | 女  |   26 | 444444   | NULL  |
| 11 | 大梅  | 女  |   27 | 555555   | NULL  |
+----+-------+-----+------+----------+-------+
5 rows in set (0.00 sec)
```

#查询性别不为男的数据
```
mysql> SELECT * FROM user WHERE sex <> '男';
+----+-------+-----+------+----------+-------+
| id | name  | sex | age  | password | phone |
+----+-------+-----+------+----------+-------+
|  3 | david | 女  |   28 | 111111   | NULL  |
|  4 | 小红  | 女  |   27 | 123456   | NULL  |
|  8 | 小绿  | 女  |   34 | 222222   | NULL  |
| 10 | 小影  | 女  |   26 | 444444   | NULL  |
| 11 | 大梅  | 女  |   27 | 555555   | NULL  |
+----+-------+-----+------+----------+-------+
5 rows in set (0.00 sec)

mysql>
```

4.2.3 模糊查询过滤

模糊查询过滤使用关键字LIKE进行查询。

（1）LIKE '张%'：使用LIKE查询该字段以"张"开头的数据。

SELECT * FROM user name like '张%';

（2）LIKE '%明'：使用LIKE查询该字段以"明"结尾的数据。

SELECT * FROM user name like '%明';

（3）LIKE '%明%'：使用LIKE查询该字段包含"明"的数据。

SELECT * FROM user name like '%明%';

实战演练——模糊查询过滤

Microsoft Windows [版本 6.1.7601]
版权所有 (c) 2009 Microsoft Corporation。保留所有权利。

C:\Users\Administrator>mysql -u root -p123456
mysql: [Warning] Using a password on the command line interface can be insecure.

Welcome to the MySQL monitor. Commands end with ; or \g.
Your MySQL connection id is 25
Server version: 5.7.20 MySQL Community Server (GPL)

Copyright (c) 2000, 2017, Oracle and/or its affiliates. All rights reserved.

Oracle is a registered trademark of Oracle Corporation and/or its
affiliates. Other names may be trademarks of their respective
owners.

Type 'help;' or '\h' for help. Type '\c' to clear the current input statement.

mysql> USE shop;
Database changed

#查询user表里的所有数据
mysql> SELECT * FROM user;
+----+-------+-----+-----+----------+-------+
| id | name | sex | age | password | phone |
+----+-------+-----+-----+----------+-------+
3	david	女	28	111111	NULL
4	小红	女	27	123456	NULL
5	小明	男	10	123456	NULL
6	小刚	男	12	123456	NULL
7	小王	男	14	111111	NULL
8	小绿	女	34	222222	NULL
9	晓峰	男	15	333333	NULL
10	小影	女	26	444444	NULL
11	大梅	女	27	555555	NULL
+----+-------+-----+-----+----------+-------+
9 rows in set (0.00 sec)

#查询姓名以"小"开头的数据

```
mysql> SELECT * FROM user WHERE name like '小%';
+----+------+-----+-----+----------+-------+
| id | name | sex | age | password | phone |
+----+------+-----+-----+----------+-------+
|  4 | 小红 | 女  | 27  | 123456   | NULL  |
|  5 | 小明 | 男  | 10  | 123456   | NULL  |
|  6 | 小刚 | 男  | 12  | 123456   | NULL  |
|  7 | 小王 | 男  | 14  | 111111   | NULL  |
|  8 | 小绿 | 女  | 34  | 222222   | NULL  |
| 10 | 小影 | 女  | 26  | 444444   | NULL  |
+----+------+-----+-----+----------+-------+
6 rows in set (0.00 sec)
```

#查询姓名以"明"结尾的数据
```
mysql> SELECT * FROM user WHERE name like '%明';
+----+------+-----+-----+----------+-------+
| id | name | sex | age | password | phone |
+----+------+-----+-----+----------+-------+
|  5 | 小明 | 男  | 10  | 123456   | NULL  |
+----+------+-----+-----+----------+-------+
1 row in set (0.00 sec)
```

#查询姓名包含"小"的数据
```
mysql> SELECT * FROM user WHERE name like '%小%';
+----+------+-----+-----+----------+-------+
| id | name | sex | age | password | phone |
+----+------+-----+-----+----------+-------+
|  4 | 小红 | 女  | 27  | 123456   | NULL  |
|  5 | 小明 | 男  | 10  | 123456   | NULL  |
|  6 | 小刚 | 男  | 12  | 123456   | NULL  |
|  7 | 小王 | 男  | 14  | 111111   | NULL  |
|  8 | 小绿 | 女  | 34  | 222222   | NULL  |
| 10 | 小影 | 女  | 26  | 444444   | NULL  |
+----+------+-----+-----+----------+-------+
6 rows in set (0.00 sec)

mysql>
```

4.2.4 字段控制查询过滤

字段控制查询过滤可以使用DISTINCT去除重复过滤，使用AS设置别名。

（1）DISTINCT：去除重复的列值。

SELECT distinct age FROM user;

（2）AS：可以设置列的别名，也可以省略AS来设置关键字。

SELECT name as 姓名 FROM user;
或者
SELECT name 姓名 FROM user;

实战演练——字段控制查询过滤

Microsoft Windows [版本 6.1.7601]
版权所有 (c) 2009 Microsoft Corporation。保留所有权利。

```
C:\Users\Administrator>mysql -u root -p123456
mysql: [Warning] Using a password on the command line interface can be insecure.

Welcome to the MySQL monitor.  Commands end with ; or \g.
Your MySQL connection id is 26
Server version: 5.7.20 MySQL Community Server (GPL)

Copyright (c) 2000, 2017, Oracle and/or its affiliates. All rights reserved.

Oracle is a registered trademark of Oracle Corporation and/or its
affiliates. Other names may be trademarks of their respective
owners.

Type 'help;' or '\h' for help. Type '\c' to clear the current input statement.

mysql> USE shop;
Database changed

#查询user表里的所有数据
mysql> SELECT * FROM user;
+----+-------+-----+-----+----------+-------+
| id | name  | sex | age | password | phone |
+----+-------+-----+-----+----------+-------+
|  3 | david | 女  |  28 | 111111   | NULL  |
|  4 | 小红  | 女  |  27 | 123456   | NULL  |
|  5 | 小明  | 男  |  10 | 123456   | NULL  |
|  6 | 小刚  | 男  |  12 | 123456   | NULL  |
|  7 | 小王  | 男  |  14 | 111111   | NULL  |
|  8 | 小绿  | 女  |  34 | 222222   | NULL  |
|  9 | 晓峰  | 男  |  15 | 333333   | NULL  |
| 10 | 小影  | 女  |  26 | 444444   | NULL  |
| 11 | 大梅  | 女  |  27 | 555555   | NULL  |
+----+-------+-----+-----+----------+-------+
9 rows in set (0.00 sec)

#去重查询性别
mysql> SELECT distinct sex FROM user;
+-----+
| sex |
+-----+
| 女  |
| 男  |
+-----+
2 rows in set (0.00 sec)

#去重查询年龄
mysql> SELECT distinct age FROM user;
+-----+
| age |
+-----+
|  28 |
```

```
|  27 |
|  10 |
|  12 |
|  14 |
|  34 |
|  15 |
|  26 |
+-----+
8 rows in set (0.00 sec)
```

#查询id与age之和
```
mysql> SELECT *,id+age FROM user;
+----+-------+-----+-----+----------+-------+--------+
| id | name  | sex | age | password | phone | id+age |
+----+-------+-----+-----+----------+-------+--------+
|  3 | david | 女  |  28 | 111111   | NULL  |     31 |
|  4 | 小红  | 女  |  27 | 123456   | NULL  |     31 |
|  5 | 小明  | 男  |  10 | 123456   | NULL  |     15 |
|  6 | 小刚  | 男  |  12 | 123456   | NULL  |     18 |
|  7 | 小王  | 男  |  14 | 111111   | NULL  |     21 |
|  8 | 小绿  | 女  |  34 | 222222   | NULL  |     42 |
|  9 | 晓峰  | 男  |  15 | 333333   | NULL  |     24 |
| 10 | 小影  | 女  |  26 | 444444   | NULL  |     36 |
| 11 | 大梅  | 女  |  27 | 555555   | NULL  |     38 |
+----+-------+-----+-----+----------+-------+--------+
9 rows in set (0.00 sec)
```

#查询id与age之和,使用AS关键字设置别名total
```
mysql> SELECT *,id+age as total FROM user;
+----+-------+-----+-----+----------+-------+-------+
| id | name  | sex | age | password | phone | total |
+----+-------+-----+-----+----------+-------+-------+
|  3 | david | 女  |  28 | 111111   | NULL  |    31 |
|  4 | 小红  | 女  |  27 | 123456   | NULL  |    31 |
|  5 | 小明  | 男  |  10 | 123456   | NULL  |    15 |
|  6 | 小刚  | 男  |  12 | 123456   | NULL  |    18 |
|  7 | 小王  | 男  |  14 | 111111   | NULL  |    21 |
|  8 | 小绿  | 女  |  34 | 222222   | NULL  |    42 |
|  9 | 晓峰  | 男  |  15 | 333333   | NULL  |    24 |
| 10 | 小影  | 女  |  26 | 444444   | NULL  |    36 |
| 11 | 大梅  | 女  |  27 | 555555   | NULL  |    38 |
+----+-------+-----+-----+----------+-------+-------+
9 rows in set (0.00 sec)
```

#查询id与age之和,不使用AS关键字设置别名total
```
mysql> SELECT *,id+age total FROM user;
+----+-------+-----+-----+----------+-------+-------+
| id | name  | sex | age | password | phone | total |
+----+-------+-----+-----+----------+-------+-------+
|  3 | david | 女  |  28 | 111111   | NULL  |    31 |
|  4 | 小红  | 女  |  27 | 123456   | NULL  |    31 |
|  5 | 小明  | 男  |  10 | 123456   | NULL  |    15 |
```

```
|  6 | 小刚  | 男  | 12 | 123456  | NULL | 18 |
|  7 | 小王  | 男  | 14 | 111111  | NULL | 21 |
|  8 | 小绿  | 女  | 34 | 222222  | NULL | 42 |
|  9 | 晓峰  | 男  | 15 | 333333  | NULL | 24 |
| 10 | 小影  | 女  | 26 | 444444  | NULL | 36 |
| 11 | 大梅  | 女  | 27 | 555555  | NULL | 38 |
+----+------+-----+----+---------+------+----+
9 rows in set (0.00 sec)

mysql>
```

4.2.5 正则表达式查询过滤

正则表达式用来匹配文本中的特殊字符串或字符集合。将正则表达式与一个文本串进行比较，查询出满足正则表达式的数据。所有种类的程序设计语言、文本编辑器、操作系统等都支持正则表达式；MySQL用WHERE子句对正则表达式提供了初步的支持，使用REGEXP关键字指定正则表达式的字符匹配模式，允许指定用正则表达式过滤SELECT检索出的数据。

在使用MySQL进行正则表达式查询之前，要掌握特殊字符在正则表达式里的使用，如表4.1所示。

表4.1 正则表达式特殊字符的使用

匹配模式	含 义	例子说明	示 例
^	匹配文本开始字符	^b 匹配以字母b开头的字符串	banner,bag
$	匹配文本结束字符	st$ 匹配以st结尾的字符串	test,persist
.	匹配任何单个字符	b.t 匹配任何b和t之间有一个字符	bit,bat,but
*	匹配0个或多个在它前面的字符	f*n 匹配字符n前面有任意个字符	fn,fan,faan
+	匹配前面的字符1次或多次	ba+ 匹配以b开头后面紧跟至少一个a	ba,bay,bare
<字符串>	匹配包含指定字符串的文本	fa	fan,afa,faad
[字符集合]	匹配字符集合中的任何一个字符	[xz] 匹配x或者z	dizzy,zebra,x-ray
[^]	匹配不在括号中的任何字符	[^abc] 匹配任何不包含a,b,c的字符串	desk,fox,f8ke
字符串{n,}	匹配前面的字符串至少n次	b{2} 匹配2个或更多b	bbb,bbb,bbbbbb
字符串{n,m}	匹配前面的字符串至少n次，至多m次。如果n为0，此参数为可选参数	b{2,4} 匹配最少2个n，最多4个b	bb,bbb,bbbb

下面来完成查询以特定字符或字符串开头的记录、查询以特定字符或字符串结尾的记录、用符号"."来替代字符串中的任意一个字符、使用"*"和"+"来匹配多个字符、匹配指定字符串、匹配指定字符中的任意一个、匹配指定字符以外的字符、使用{n,}或者{n,m}来指定字符串连续出现的次数。

实战演练 —— 使用正则表达式

Microsoft Windows [版本 6.1.7601]
版权所有 (c) 2009 Microsoft Corporation。保留所有权利。

C:\Users\Administrator>mysql -u root -p123456
mysql: [Warning] Using a password on the command line interface can be insecure.

Welcome to the MySQL monitor. Commands end with ; or \g.
Your MySQL connection id is 30

Server version: 5.7.20 MySQL Community Server (GPL)

Copyright (c) 2000, 2017, Oracle and/or its affiliates. All rights reserved.

Oracle is a registered trademark of Oracle Corporation and/or its
affiliates. Other names may be trademarks of their respective
owners.

Type 'help;' or '\h' for help. Type '\c' to clear the current input statement.

mysql> USE shop;
Database changed

#查询user表里的所有记录
mysql> SELECT * FROM user;

```
+----+------+-----+-----+----------+-------+-----------+--------+
| id | name | sex | age | password | phone | loginName | remark |
+----+------+-----+-----+----------+-------+-----------+--------+
|  3 | david| 女  |  28 | 111111   | NULL  | david     | baann  |
|  4 | 小红 | 女  |  27 | 123456   | NULL  | xiaohong  | black  |
|  5 | 小明 | 男  |  10 | 123456   | NULL  | xiaoming  | berry  |
|  6 | 小刚 | 男  |  12 | 123456   | NULL  | xiaogang  | banner |
|  7 | 小王 | 男  |  14 | 111111   | NULL  | xiaowang  | banana |
|  8 | 小绿 | 女  |  34 | 222222   | NULL  | xiaolv    | car    |
|  9 | 晓峰 | 男  |  15 | 333333   | NULL  | xiaofeng  | carray |
| 10 | 小影 | 女  |  26 | 444444   | NULL  | xiaoying  | baaaa  |
| 11 | 大梅 | 女  |  27 | 555555   | NULL  | damei     | accc   |
+----+------+-----+-----+----------+-------+-----------+--------+
```
9 rows in set (0.00 sec)

#查询remark字段里以b开头的数据
mysql> SELECT * FROM user WHERE remark regexp '^b';

```
+----+------+-----+-----+----------+-------+-----------+--------+
| id | name | sex | age | password | phone | loginName | remark |
+----+------+-----+-----+----------+-------+-----------+--------+
|  3 | david| 女  |  28 | 111111   | NULL  | david     | baann  |
|  4 | 小红 | 女  |  27 | 123456   | NULL  | xiaohong  | black  |
|  5 | 小明 | 男  |  10 | 123456   | NULL  | xiaoming  | berry  |
|  6 | 小刚 | 男  |  12 | 123456   | NULL  | xiaogang  | banner |
|  7 | 小王 | 男  |  14 | 111111   | NULL  | xiaowang  | banana |
| 10 | 小影 | 女  |  26 | 444444   | NULL  | xiaoying  | baaaa  |
+----+------+-----+-----+----------+-------+-----------+--------+
```
6 rows in set (0.00 sec)

#查询loginName字段里以ang结尾的数据
mysql> SELECT * FROM user WHERE loginName regexp 'ang$';

```
+----+------+-----+-----+----------+-------+-----------+--------+
| id | name | sex | age | password | phone | loginName | remark |
+----+------+-----+-----+----------+-------+-----------+--------+
|  6 | 小刚 | 男  |  12 | 123456   | NULL  | xiaogang  | banner |
|  7 | 小王 | 男  |  14 | 111111   | NULL  | xiaowang  | banana |
+----+------+-----+-----+----------+-------+-----------+--------+
```

2 rows in set (0.00 sec)

#用符号"."来替代字符串中的任意一个字符
```
mysql> SELECT * FROM user WHERE loginName regexp 'x.a';
+----+------+-----+-----+----------+-------+-----------+--------+
| id | name | sex | age | password | phone | loginName | remark |
+----+------+-----+-----+----------+-------+-----------+--------+
|  4 | 小红 | 女  | 27  | 123456   | NULL  | xiaohong  | black  |
|  5 | 小明 | 男  | 10  | 123456   | NULL  | xiaoming  | berry  |
|  6 | 小刚 | 男  | 12  | 123456   | NULL  | xiaogang  | banner |
|  7 | 小王 | 男  | 14  | 111111   | NULL  | xiaowang  | banana |
|  8 | 小绿 | 女  | 34  | 222222   | NULL  | xiaolv    | car    |
|  9 | 晓峰 | 男  | 15  | 333333   | NULL  | xiaofeng  | carray |
| 10 | 小影 | 女  | 26  | 444444   | NULL  | xiaoying  | baaaa  |
+----+------+-----+-----+----------+-------+-----------+--------+
```
7 rows in set (0.00 sec)

#用星号'*'匹配前面的字符任意次，包括0次
```
mysql> SELECT * FROM user WHERE remark regexp '^ba*';
+----+-------+-----+-----+----------+-------+-----------+--------+
| id | name  | sex | age | password | phone | loginName | remark |
+----+-------+-----+-----+----------+-------+-----------+--------+
|  3 | david | 女  | 28  | 111111   | NULL  | david     | baann  |
|  4 | 小红  | 女  | 27  | 123456   | NULL  | xiaohong  | black  |
|  5 | 小明  | 男  | 10  | 123456   | NULL  | xiaoming  | berry  |
|  6 | 小刚  | 男  | 12  | 123456   | NULL  | xiaogang  | banner |
|  7 | 小王  | 男  | 14  | 111111   | NULL  | xiaowang  | banana |
| 10 | 小影  | 女  | 26  | 444444   | NULL  | xiaoying  | baaaa  |
+----+-------+-----+-----+----------+-------+-----------+--------+
```
6 rows in set (0.00 sec)

#用加号'+'匹配前面的字符至少一次
```
mysql> SELECT * FROM user WHERE remark regexp '^ba+';
+----+-------+-----+-----+----------+-------+-----------+--------+
| id | name  | sex | age | password | phone | loginName | remark |
+----+-------+-----+-----+----------+-------+-----------+--------+
|  3 | david | 女  | 28  | 111111   | NULL  | david     | baann  |
|  6 | 小刚  | 男  | 12  | 123456   | NULL  | xiaogang  | banner |
|  7 | 小王  | 男  | 14  | 111111   | NULL  | xiaowang  | banana |
| 10 | 小影  | 女  | 26  | 444444   | NULL  | xiaoying  | baaaa  |
+----+-------+-----+-----+----------+-------+-----------+--------+
```
4 rows in set (0.00 sec)

#匹配指定字符串
```
mysql> SELECT * FROM user WHERE loginName regexp 'xiao';
+----+------+-----+-----+----------+-------+-----------+--------+
| id | name | sex | age | password | phone | loginName | remark |
+----+------+-----+-----+----------+-------+-----------+--------+
|  4 | 小红 | 女  | 27  | 123456   | NULL  | xiaohong  | black  |
|  5 | 小明 | 男  | 10  | 123456   | NULL  | xiaoming  | berry  |
|  6 | 小刚 | 男  | 12  | 123456   | NULL  | xiaogang  | banner |
|  7 | 小王 | 男  | 14  | 111111   | NULL  | xiaowang  | banana |
```

```
|  8 | 小绿  | 女  | 34 | 222222 | NULL | xiaolv   | car    |
|  9 | 晓峰  | 男  | 15 | 333333 | NULL | xiaofeng | carray |
| 10 | 小影  | 女  | 26 | 444444 | NULL | xiaoying | baaaa  |
+----+-------+-----+----+--------+------+----------+--------+
```
7 rows in set (0.00 sec)

#匹配多个字符串，多个字符串之间使用分隔符'|'隔开
mysql> SELECT * FROM user WHERE loginName regexp 'xiao|mei';
```
+----+-------+-----+-----+----------+-------+-----------+---------+
| id | name  | sex | age | password | phone | loginName | remark  |
+----+-------+-----+-----+----------+-------+-----------+---------+
|  4 | 小红  | 女  | 27  | 123456   | NULL  | xiaohong  | black   |
|  5 | 小明  | 男  | 10  | 123456   | NULL  | xiaoming  | berry   |
|  6 | 小刚  | 男  | 12  | 123456   | NULL  | xiaogang  | banner  |
|  7 | 小王  | 男  | 14  | 111111   | NULL  | xiaowang  | banana  |
|  8 | 小绿  | 女  | 34  | 222222   | NULL  | xiaolv    | car     |
|  9 | 晓峰  | 男  | 15  | 333333   | NULL  | xiaofeng  | carray  |
| 10 | 小影  | 女  | 26  | 444444   | NULL  | xiaoying  | baaaa   |
| 11 | 大梅  | 女  | 27  | 555555   | NULL  | damei     | accc    |
+----+-------+-----+-----+----------+-------+-----------+---------+
```
8 rows in set (0.00 sec)

#方括号"[]"指定一个字符集合，只匹配其中任何一个字符，即为所查找的文本，还可以指定数值集合。"[a-z]"表示集合区间为a~z的字母，"[0-9]"表示集合区间为所有数字
mysql> SELECT * FROM user WHERE loginName regexp '[da]';
```
+----+-------+-----+-----+----------+-------+-----------+---------+
| id | name  | sex | age | password | phone | loginName | remark  |
+----+-------+-----+-----+----------+-------+-----------+---------+
|  3 | david | 女  | 28  | 111111   | NULL  | david     | baann   |
|  4 | 小红  | 女  | 27  | 123456   | NULL  | xiaohong  | black   |
|  5 | 小明  | 男  | 10  | 123456   | NULL  | xiaoming  | berry   |
|  6 | 小刚  | 男  | 12  | 123456   | NULL  | xiaogang  | banner  |
|  7 | 小王  | 男  | 14  | 111111   | NULL  | xiaowang  | banana  |
|  8 | 小绿  | 女  | 34  | 222222   | NULL  | xiaolv    | car     |
|  9 | 晓峰  | 男  | 15  | 333333   | NULL  | xiaofeng  | carray  |
| 10 | 小影  | 女  | 26  | 444444   | NULL  | xiaoying  | baaaa   |
| 11 | 大梅  | 女  | 27  | 555555   | NULL  | damei     | accc    |
+----+-------+-----+-----+----------+-------+-----------+---------+
```
9 rows in set (0.00 sec)

#查询不在这个集合范围内的数据
mysql> SELECT * FROM user WHERE id regexp '[^1-4]';
```
+----+-------+-----+-----+----------+-------+-----------+---------+
| id | name  | sex | age | password | phone | loginName | remark  |
+----+-------+-----+-----+----------+-------+-----------+---------+
|  5 | 小明  | 男  | 10  | 123456   | NULL  | xiaoming  | berry   |
|  6 | 小刚  | 男  | 12  | 123456   | NULL  | xiaogang  | banner  |
|  7 | 小王  | 男  | 14  | 111111   | NULL  | xiaowang  | banana  |
|  8 | 小绿  | 女  | 34  | 222222   | NULL  | xiaolv    | car     |
|  9 | 晓峰  | 男  | 15  | 333333   | NULL  | xiaofeng  | carray  |
| 10 | 小影  | 女  | 26  | 444444   | NULL  | xiaoying  | baaaa   |
+----+-------+-----+-----+----------+-------+-----------+---------+
```
6 rows in set (0.00 sec)

```
#字符串{n,}：匹配前面的字符串至少n次
mysql> SELECT * FROM user WHERE remark regexp 'a{2,}';
+----+------+-----+-----+--------+-------+-----------+--------+
| id | name | sex | age | password | phone | loginName | remark |
+----+------+-----+-----+--------+-------+-----------+--------+
| 3  | david| 女  | 28  | 111111 | NULL  | david     | baann  |
| 10 | 小影 | 女  | 26  | 444444 | NULL  | xiaoying  | baaaa  |
+----+------+-----+-----+--------+-------+-----------+--------+
2 rows in set (0.00 sec)

#字符串{n,m}：匹配前面的字符串至少n次，至多m次。如果n为0，此参数为可选参数
mysql> SELECT * FROM user WHERE remark regexp 'ba{1,3}';
+----+------+-----+-----+--------+-------+-----------+--------+
| id | name | sex | age | password | phone | loginName | remark |
+----+------+-----+-----+--------+-------+-----------+--------+
| 3  | david| 女  | 28  | 111111 | NULL  | david     | baann  |
| 6  | 小刚 | 男  | 12  | 123456 | NULL  | xiaogang  | banner |
| 7  | 小王 | 男  | 14  | 111111 | NULL  | xiaowang  | banana |
| 10 | 小影 | 女  | 26  | 444444 | NULL  | xiaoying  | baaaa  |
+----+------+-----+-----+--------+-------+-----------+--------+
4 rows in set (0.00 sec)

mysql>
```

4.3 子查询

4.3.1 什么是子查询

子查询是一个父表达式调用另一个子表达式结果的查询操作，子表达式结果传递给父表达式继续处理，子查询也被称为内嵌查询或者内部查询。子查询可以包含普通SELECT可以包括的任何子句，如distinct、GROUP BY、ORDEY BY、LIMIT、JOIN和UNION等；但是对应的外部查询必须是以下语句之一：SELECT、INSERT、UPDATE、DELETE、set或者do。

子查询可以按返回结果和按对返回结果的调用方法进行分类。按返回结果分类的子查询又可以分为表子查询、行子查询、列子查询、标量子查询；按对返回结果的调用方法分类的子查询可以分为WHERE子查询、FROM子查询、exists子查询。

精讲视频

子查询

4.3.2 按返回结果分类的子查询

按返回结果分类的子查询又可以分为表子查询、行子查询、列子查询、标量子查询4种。

（1）表子查询：返回的结果集由多行（至少一行）数据组成，作为表子查询要设置表的别名，常用于父查询的FROM子句中。

查询用户表中年龄大于20岁的用户的姓名、性别、年龄。

SELECT * FROM (SELECT name,sex,age FROM user WHERE age >20) as user20;

实战演练——表子查询

Microsoft Windows [版本 6.1.7601]
版权所有 (c) 2009 Microsoft Corporation。保留所有权利。

```
C:\Users\Administrator>mysql -u root -p123456
mysql: [Warning] Using a password on the command line interface can be insecure.

Welcome to the MySQL monitor.  Commands end with ; or \g.
Your MySQL connection id is 151
Server version: 5.7.20 MySQL Community Server (GPL)

Copyright (c) 2000, 2017, Oracle and/or its affiliates. All rights reserved.

Oracle is a registered trademark of Oracle Corporation and/or its
affiliates. Other names may be trademarks of their respective
owners.

Type 'help;' or '\h' for help. Type '\c' to clear the current input statement.

mysql> USE shop;
Database changed

#查询用户表中年龄大于20岁的用户的姓名、性别、年龄
mysql> SELECT * FROM (SELECT name,sex,age FROM user WHERE age > 20) as user20;
+------+-----+-----+
| name | sex | age |
+------+-----+-----+
| david | 女  | 28  |
| 小红  | 女  | 27  |
| 小绿  | 女  | 34  |
| 小影  | 女  | 26  |
| 大梅  | 女  | 27  |
+------+-----+-----+
5 rows in set (0.09 sec)

mysql>
```

（2）行子查询：返回结果集由一行数据组成，一行数据库里可以包含多列数据，常用于父查询的FROM子句中或者WHERE子句中。

查询和david一样性别的用户。

SELECT * FROM user WHERE sex = (SELECT sex FROM user WHERE name='david');

实战演练——行子查询

```
Microsoft Windows [版本 6.1.7601]
版权所有 (c) 2009 Microsoft Corporation。保留所有权利。

C:\Users\Administrator>mysql -u root -p123456
mysql: [Warning] Using a password on the command line interface can be insecure.

Welcome to the MySQL monitor.  Commands end with ; or \g.
Your MySQL connection id is 151
Server version: 5.7.20 MySQL Community Server (GPL)

Copyright (c) 2000, 2017, Oracle and/or its affiliates. All rights reserved.
```

Oracle is a registered trademark of Oracle Corporation and/or its
affiliates. Other names may be trademarks of their respective
owners.

Type 'help;' or '\h' for help. Type '\c' to clear the current input statement.

mysql> USE shop;
Database changed

#查询和david一样性别的用户
mysql> SELECT * FROM user WHERE sex = (SELECT sex FROM user WHERE name='david');

```
+----+------+------+------+----------+-------+-----------+--------+
| id | name | sex  | age  | password | phone | loginName | remark |
+----+------+------+------+----------+-------+-----------+--------+
|  3 | david| 女   |  28  | 111111   | NULL  | david     | baann  |
|  4 | 小红 | 女   |  27  | 123456   | NULL  | xiaohong  | black  |
|  8 | 小绿 | 女   |  34  | 222222   | NULL  | xiaolv    | car    |
| 10 | 小影 | 女   |  26  | 444444   | NULL  | xiaoying  | baaaa  |
| 11 | 大梅 | 女   |  27  | 555555   | NULL  | damei     | accc   |
+----+------+------+------+----------+-------+-----------+--------+
```
5 rows in set (0.06 sec)

mysql>

（3）列子查询：返回的结果集由多行一列数据组成，可以使用 in、any和all操作符。

in在指定项内，同 IN(项1,项2,…)；any与比较操作符联合使用，any关键字必须接在一个比较操作符的后面，表示与子查询返回的任何值比较为true，则返回 true；all与比较操作符联合使用，all 关键字必须接在一个比较操作符的后面，表示与子查询返回的所有值比较都为true，则返回 true。

查询性别为女的id和name。

SELECT id,name FROM user WHERE id in (SELECT id FROM user WHERE sex='女') ;

查询年龄大于david的用户的id、name、age。

SELECT id,name,age FROM user WHERE age > any (SELECT age FROM user WHERE name='david');

查询年龄最小的用户。

SELECT * FROM user WHERE age <= all (SELECT age FROM user);

实战演练——列子查询

Microsoft Windows [版本 6.1.7601]
版权所有 (c) 2009 Microsoft Corporation。保留所有权利。

C:\Users\Administrator>mysql -u root -p123456
mysql: [Warning] Using a password on the command line interface can be insecure.
Welcome to the MySQL monitor. Commands end with ; or \g.
Your MySQL connection id is 152
Server version: 5.7.20 MySQL Community Server (GPL)

Copyright (c) 2000, 2017, Oracle and/or its affiliates. All rights reserved.

```
Oracle is a registered trademark of Oracle Corporation and/or its
affiliates. Other names may be trademarks of their respective
owners.

Type 'help;' or '\h' for help. Type '\c' to clear the current input statement.
```

mysql> USE shop;
Database changed

mysql> SELECT * FROM user;
```
+----+------+-----+-----+----------+-------+-----------+--------+
| id | name | sex | age | password | phone | loginName | remark |
+----+------+-----+-----+----------+-------+-----------+--------+
|  3 | david| 女  | 28  | 111111   | NULL  | david     | baann  |
|  4 | 小红 | 女  | 27  | 123456   | NULL  | xiaohong  | black  |
|  5 | 小明 | 男  | 10  | 123456   | NULL  | xiaoming  | berry  |
|  6 | 小刚 | 男  | 12  | 123456   | NULL  | xiaogang  | banner |
|  7 | 小王 | 男  | 14  | 111111   | NULL  | xiaowang  | banana |
|  8 | 小绿 | 女  | 34  | 222222   | NULL  | xiaolv    | car    |
|  9 | 晓峰 | 男  | 15  | 333333   | NULL  | xiaofeng  | carray |
| 10 | 小影 | 女  | 26  | 444444   | NULL  | xiaoying  | baaaa  |
| 11 | 大梅 | 女  | 27  | 555555   | NULL  | damei     | accc   |
+----+------+-----+-----+----------+-------+-----------+--------+
9 rows in set (0.01 sec)
```

#查询性别为女的id和name
mysql> SELECT id,name FROM user WHERE id in (SELECT id FROM user WHERE sex='女') ;
```
+----+------+
| id | name |
+----+------+
|  3 | david|
|  4 | 小红 |
|  8 | 小绿 |
| 10 | 小影 |
| 11 | 大梅 |
+----+------+
5 rows in set (0.00 sec)
```

#查询年龄大于david的用户的id、name、age
mysql> SELECT id,name,age FROM user WHERE age > any (SELECT age FROM user WHERE name='david');
```
+----+------+-----+
| id | name | age |
+----+------+-----+
|  8 | 小绿 | 34  |
+----+------+-----+
1 row in set (0.00 sec)
```

#查询年龄最小的用户
mysql> SELECT * FROM user WHERE age <= all (SELECT age FROM user);
```
+----+------+-----+-----+----------+-------+-----------+--------+
| id | name | sex | age | password | phone | loginName | remark |
```

```
+----+------+------+------+----------+-------+-----------+--------+
| 5  | 小明 | 男   | 10   | 123456   | NULL  | xiaoming  | berry  |
+----+------+------+------+----------+-------+-----------+--------+
1 row in set (0.00 sec)

mysql>
```

（4）标量子查询：返回的结果集是一个标量集合，一行一列，也就是一个标量值；每个标量子查询也是一个行子查询和一个列子查询，反之则不是；每个行子查询和列子查询也是一个表子查询，反之则不是。

查询和id等于4一样年龄的用户。

SELECT name FROM user WHERE age = (SELECT age FROM user WHERE id=4) and id !=4;

实战演练——标量子查询

Microsoft Windows [版本 6.1.7601]
版权所有 (c) 2009 Microsoft Corporation。保留所有权利。

C:\Users\Administrator>mysql –u root –p123456
mysql: [Warning] Using a password on the command line interface can be insecure.

Welcome to the MySQL monitor. Commands end with ; or \g.
Your MySQL connection id is 153
Server version: 5.7.20 MySQL Community Server (GPL)

Copyright (c) 2000, 2017, Oracle and/or its affiliates. All rights reserved.

Oracle is a registered trademark of Oracle Corporation and/or its
affiliates. Other names may be trademarks of their respective
owners.

Type 'help;' or '\h' for help. Type '\c' to clear the current input statement.

mysql> USE shop;
Database changed

mysql> SELECT * FROM user;

```
+----+--------+------+------+----------+-------+-----------+--------+
| id | name   | sex  | age  | password | phone | loginName | remark |
+----+--------+------+------+----------+-------+-----------+--------+
| 3  | david  | 女   | 28   | 111111   | NULL  | david     | baann  |
| 4  | 小红   | 女   | 27   | 123456   | NULL  | xiaohong  | black  |
| 5  | 小明   | 男   | 10   | 123456   | NULL  | xiaoming  | berry  |
| 6  | 小刚   | 男   | 12   | 123456   | NULL  | xiaogang  | banner |
| 7  | 小王   | 男   | 14   | 111111   | NULL  | xiaowang  | banana |
| 8  | 小绿   | 女   | 34   | 222222   | NULL  | xiaolv    | car    |
| 9  | 晓峰   | 男   | 15   | 333333   | NULL  | xiaofeng  | carray |
| 10 | 小影   | 女   | 26   | 444444   | NULL  | xiaoying  | baaaa  |
| 11 | 大梅   | 女   | 27   | 555555   | NULL  | damei     | accc   |
+----+--------+------+------+----------+-------+-----------+--------+
9 rows in set (0.00 sec)
```

#查询和id等于4一样年龄的用户

```
mysql> SELECT name FROM user WHERE age = (SELECT age FROM user WHERE id=4) and id !=4;
+--------+
| name   |
+--------+
| 大梅   |
+--------+
1 row in set (0.07 sec)

mysql>
```

4.3.3 按对返回结果的调用方法分类的子查询

按对返回结果的调用方法分类的子查询又可以分为WHERE型子查询、FROM型子查询、exists型子查询。

（1）WHERE型子查询：把内层查询的结果作为外层查询的条件。

查询比小刚年龄大的用户。

SELECT * FROM user WHERE age > (SELECT age FROM user WHERE name='小刚');

实战演练 ——WHERE型子查询

```
Microsoft Windows [版本 6.1.7601]
版权所有 (c) 2009 Microsoft Corporation。保留所有权利。

C:\Users\Administrator>mysql –u root –p123456
mysql: [Warning] Using a password on the command line interface can be insecure.

Welcome to the MySQL monitor.  Commands end with ; or \g.
Your MySQL connection id is 154
Server version: 5.7.20 MySQL Community Server (GPL)

Copyright (c) 2000, 2017, Oracle and/or its affiliates. All rights reserved.

Oracle is a registered trademark of Oracle Corporation and/or its
affiliates. Other names may be trademarks of their respective
owners.

Type 'help;' or '\h' for help. Type '\c' to clear the current input statement.

mysql> USE shop;
Database changed

mysql> SELECT * FROM user;
+----+-------+-----+-----+----------+-------+-----------+--------+
| id | name  | sex | age | password | phone | loginName | remark |
+----+-------+-----+-----+----------+-------+-----------+--------+
|  3 | david | 女  |  28 | 111111   | NULL  | david     | baann  |
|  4 | 小红  | 女  |  27 | 123456   | NULL  | xiaohong  | black  |
|  5 | 小明  | 男  |  10 | 123456   | NULL  | xiaoming  | berry  |
|  6 | 小刚  | 男  |  12 | 123456   | NULL  | xiaogang  | banner |
|  7 | 小王  | 男  |  14 | 111111   | NULL  | xiaowang  | banana |
|  8 | 小绿  | 女  |  34 | 222222   | NULL  | xiaolv    | car    |
|  9 | 晓峰  | 男  |  15 | 333333   | NULL  | xiaofeng  | carray |
| 10 | 小影  | 女  |  26 | 444444   | NULL  | xiaoying  | baaaa  |
```

```
| 11 | 大梅  | 女  | 27 | 555555    | NULL  | damei     | accc   |
+----+-------+-----+----+-----------+-------+-----------+--------+
9 rows in set (0.00 sec)
```

#查询比小刚年龄大的用户
```
mysql> SELECT * FROM user WHERE age > (SELECT age FROM user WHERE name='小刚');
+----+-------+-----+----+-----------+-------+-----------+--------+
| id | name  | sex | age| password  | phone | loginName | remark |
+----+-------+-----+----+-----------+-------+-----------+--------+
|  3 | david | 女  | 28 | 111111    | NULL  | david     | baann  |
|  4 | 小红  | 女  | 27 | 123456    | NULL  | xiaohong  | black  |
|  7 | 小王  | 男  | 14 | 111111    | NULL  | xiaowang  | banana |
|  8 | 小绿  | 女  | 34 | 222222    | NULL  | xiaolv    | car    |
|  9 | 晓峰  | 男  | 15 | 333333    | NULL  | xiaofeng  | carray |
| 10 | 小影  | 女  | 26 | 444444    | NULL  | xiaoying  | baaaa  |
| 11 | 大梅  | 女  | 27 | 555555    | NULL  | damei     | accc   |
+----+-------+-----+----+-----------+-------+-----------+--------+
7 rows in set (0.05 sec)

mysql>
```

（2）FROM型子查询：将返回的结果集作为一个临时表，临时表要设置别名，然后在临时表中进行查询。
查询性别为男并且年龄大于12岁的用户。

SELECT * FROM (SELECT name,loginName,age,password FROM user WHERE sex='男') as temp WHERE temp.age > 12 ;

实战演练——FROM型子查询

Microsoft Windows [版本 6.1.7601]
版权所有 (c) 2009 Microsoft Corporation。保留所有权利。

C:\Users\Administrator>mysql –u root –p123456
mysql: [Warning] Using a password on the command line interface can be insecure.

Welcome to the MySQL monitor. Commands end with ; or \g.
Your MySQL connection id is 154
Server version: 5.7.20 MySQL Community Server (GPL)

Copyright (c) 2000, 2017, Oracle and/or its affiliates. All rights reserved.

Oracle is a registered trademark of Oracle Corporation and/or its
affiliates. Other names may be trademarks of their respective
owners.

Type 'help;' or '\h' for help. Type '\c' to clear the current input statement.

mysql> USE shop;
Database changed

mysql> SELECT * FROM user;
```
+----+------+-----+-----+----------+-------+-----------+--------+
| id | name | sex | age | password | phone | loginName | remark |
+----+------+-----+-----+----------+-------+-----------+--------+
```

```
|  3 | david |  女 | 28 | 111111 | NULL | david    | baann  |
|  4 | 小红  |  女 | 27 | 123456 | NULL | xiaohong | black  |
|  5 | 小明  |  男 | 10 | 123456 | NULL | xiaoming | berry  |
|  6 | 小刚  |  男 | 12 | 123456 | NULL | xiaogang | banner |
|  7 | 小王  |  男 | 14 | 111111 | NULL | xiaowang | banana |
|  8 | 小绿  |  女 | 34 | 222222 | NULL | xiaolv   | car    |
|  9 | 晓峰  |  男 | 15 | 333333 | NULL | xiaofeng | carray |
| 10 | 小影  |  女 | 26 | 444444 | NULL | xiaoying | baaaa  |
| 11 | 大梅  |  女 | 27 | 555555 | NULL | damei    | accc   |
+----+-------+-----+----+--------+------+----------+--------+
```
9 rows in set (0.00 sec)

#查询性别为男的用户
mysql> SELECT * FROM (SELECT name,loginName,age,password FROM user WHERE sex='男') as temp ;

```
+--------+-----------+-----+----------+
| name   | loginName | age | password |
+--------+-----------+-----+----------+
| 小明   | xiaoming  | 10  | 123456   |
| 小刚   | xiaogang  | 12  | 123456   |
| 小王   | xiaowang  | 14  | 111111   |
| 晓峰   | xiaofeng  | 15  | 333333   |
+--------+-----------+-----+----------+
```
4 rows in set (0.00 sec)

#查询性别为男并且年龄大于12岁的用户
mysql> SELECT * FROM (SELECT name,loginName,age,password FROM user WHERE sex='男') as temp WHERE temp.age > 12 ;

```
+--------+-----------+-----+----------+
| name   | loginName | age | password |
+--------+-----------+-----+----------+
| 小王   | xiaowang  | 14  | 111111   |
| 晓峰   | xiaofeng  | 15  | 333333   |
+--------+-----------+-----+----------+
```
2 rows in set (0.00 sec)

mysql>

（3）exists型子查询：把外层查询结果拿到内层，看内层的查询是否成立，使用exists关键字时，内层查询语句不返回查询的记录，而是返回一个真假值，如果内层查询语句查询到满足条件的记录，只要子查询中至少返回一个值，则exists语句的值就为true，就返回true，否则返回false。当返回的值为true时，外层查询语句将进行查询，否则不进行查询。not exists刚好与之相反。

查询用户密码存在111111。
SELECT * FROM user WHERE exists (SELECT * FROM user WHERE password='111111');

实战演练——exists型子查询

Microsoft Windows [版本 6.1.7601]
版权所有 (c) 2009 Microsoft Corporation。保留所有权利。

C:\Users\Administrator>mysql –u root –p123456

```
mysql: [Warning] Using a password on the command line interface can be insecure.

Welcome to the MySQL monitor.  Commands end with ; or \g.
Your MySQL connection id is 155
Server version: 5.7.20 MySQL Community Server (GPL)

Copyright (c) 2000, 2017, Oracle and/or its affiliates. All rights reserved.

Oracle is a registered trademark of Oracle Corporation and/or its
affiliates. Other names may be trademarks of their respective
owners.

Type 'help;' or '\h' for help. Type '\c' to clear the current input statement.
```

mysql> USE shop;
Database changed

mysql> SELECT * FROM user;
```
+----+--------+-----+-----+----------+-------+-----------+---------+
| id | name   | sex | age | password | phone | loginName | remark  |
+----+--------+-----+-----+----------+-------+-----------+---------+
|  3 | david  | 女  |  28 | 111111   | NULL  | david     | baann   |
|  4 | 小红   | 女  |  27 | 123456   | NULL  | xiaohong  | black   |
|  5 | 小明   | 男  |  10 | 123456   | NULL  | xiaoming  | berry   |
|  6 | 小刚   | 男  |  12 | 123456   | NULL  | xiaogang  | banner  |
|  7 | 小王   | 男  |  14 | 111111   | NULL  | xiaowang  | banana  |
|  8 | 小绿   | 女  |  34 | 222222   | NULL  | xiaolv    | car     |
|  9 | 晓峰   | 男  |  15 | 333333   | NULL  | xiaofeng  | carray  |
| 10 | 小影   | 女  |  26 | 444444   | NULL  | xiaoying  | baaaa   |
| 11 | 大梅   | 女  |  27 | 555555   | NULL  | damei     | accc    |
+----+--------+-----+-----+----------+-------+-----------+---------+
9 rows in set (0.00 sec)
```

#查询用户密码存在111111
mysql> SELECT * FROM user WHERE exists (SELECT * FROM user WHERE password='111111');
```
+----+--------+-----+-----+----------+-------+-----------+---------+
| id | name   | sex | age | password | phone | loginName | remark  |
+----+--------+-----+-----+----------+-------+-----------+---------+
|  3 | david  | 女  |  28 | 111111   | NULL  | david     | baann   |
|  4 | 小红   | 女  |  27 | 123456   | NULL  | xiaohong  | black   |
|  5 | 小明   | 男  |  10 | 123456   | NULL  | xiaoming  | berry   |
|  6 | 小刚   | 男  |  12 | 123456   | NULL  | xiaogang  | banner  |
|  7 | 小王   | 男  |  14 | 111111   | NULL  | xiaowang  | banana  |
|  8 | 小绿   | 女  |  34 | 222222   | NULL  | xiaolv    | car     |
|  9 | 晓峰   | 男  |  15 | 333333   | NULL  | xiaofeng  | carray  |
| 10 | 小影   | 女  |  26 | 444444   | NULL  | xiaoying  | baaaa   |
| 11 | 大梅   | 女  |  27 | 555555   | NULL  | damei     | accc    |
+----+--------+-----+-----+----------+-------+-----------+---------+
9 rows in set (0.00 sec)
```

#查询用户密码存在1111121
mysql> SELECT * FROM user WHERE exists (SELECT * FROM user WHERE password='1111121');

Empty set (0.04 sec)

#查询用户密码存在1111121
```
mysql> SELECT * FROM user WHERE not exists (SELECT * FROM user WHERE password='1111121');
+----+------+-----+-----+----------+-------+-----------+--------+
| id | name | sex | age | password | phone | loginName | remark |
+----+------+-----+-----+----------+-------+-----------+--------+
|  3 | david| 女  | 28  | 111111   | NULL  | david     | baann  |
|  4 | 小红 | 女  | 27  | 123456   | NULL  | xiaohong  | black  |
|  5 | 小明 | 男  | 10  | 123456   | NULL  | xiaoming  | berry  |
|  6 | 小刚 | 男  | 12  | 123456   | NULL  | xiaogang  | banner |
|  7 | 小王 | 男  | 14  | 111111   | NULL  | xiaowang  | banana |
|  8 | 小绿 | 女  | 34  | 222222   | NULL  | xiaolv    | car    |
|  9 | 晓峰 | 男  | 15  | 333333   | NULL  | xiaofeng  | carray |
| 10 | 小影 | 女  | 26  | 444444   | NULL  | xiaoying  | baaaa  |
| 11 | 大梅 | 女  | 27  | 555555   | NULL  | damei     | accc   |
+----+------+-----+-----+----------+-------+-----------+--------+
9 rows in set (0.00 sec)
```

#查询用户密码不存在1123456
```
mysql> SELECT * FROM user WHERE not exists (SELECT * FROM user WHERE password='1123456');
+----+------+-----+-----+----------+-------+-----------+--------+
| id | name | sex | age | password | phone | loginName | remark |
+----+------+-----+-----+----------+-------+-----------+--------+
|  3 | david| 女  | 28  | 111111   | NULL  | david     | baann  |
|  4 | 小红 | 女  | 27  | 123456   | NULL  | xiaohong  | black  |
|  5 | 小明 | 男  | 10  | 123456   | NULL  | xiaoming  | berry  |
|  6 | 小刚 | 男  | 12  | 123456   | NULL  | xiaogang  | banner |
|  7 | 小王 | 男  | 14  | 111111   | NULL  | xiaowang  | banana |
|  8 | 小绿 | 女  | 34  | 222222   | NULL  | xiaolv    | car    |
|  9 | 晓峰 | 男  | 15  | 333333   | NULL  | xiaofeng  | carray |
| 10 | 小影 | 女  | 26  | 444444   | NULL  | xiaoying  | baaaa  |
| 11 | 大梅 | 女  | 27  | 555555   | NULL  | damei     | accc   |
+----+------+-----+-----+----------+-------+-----------+--------+
9 rows in set (0.00 sec)
```

#查询用户密码不存在123456
```
mysql> SELECT * FROM user WHERE not exists (SELECT * FROM user WHERE password='123456');
Empty set (0.00 sec)
```

#先判断密码是否存在123456，然后输出密码等于123456
```
mysql> SELECT * FROM user WHERE exists (SELECT * FROM user WHERE password='123456') and password='123456';
+----+------+-----+-----+----------+-------+-----------+--------+
| id | name | sex | age | password | phone | loginName | remark |
+----+------+-----+-----+----------+-------+-----------+--------+
|  4 | 小红 | 女  | 27  | 123456   | NULL  | xiaohong  | black  |
|  5 | 小明 | 男  | 10  | 123456   | NULL  | xiaoming  | berry  |
|  6 | 小刚 | 男  | 12  | 123456   | NULL  | xiaogang  | banner |
+----+------+-----+-----+----------+-------+-----------+--------+
3 rows in set (0.00 sec)

mysql>
```

4.4 聚合函数

MySQL数据库提供了一些常用的聚合函数。聚合函数是运行在行组上，计算和返回单个值的函数。AVG()函数用来返回某列的平均值，COUNT()函数用来返回某列的行数，MAX()函数用来返回某列的最大值，MIN()函数用来返回某列的最小值，SUM()函数用来返回某列之和。

精讲视频

聚合函数

4.4.1 AVG()函数

MySQL数据库中的AVG函数用来计算某列的平均值，如可以用于计算平均分数、平均薪资、平均年龄等。
SELECT avg(age) FROM user;

实战演练——AVG()函数

Microsoft Windows [版本 6.1.7601]
版权所有 (c) 2009 Microsoft Corporation。保留所有权利。

C:\Users\Administrator>mysql –u root –p123456
mysql: [Warning] Using a password on the command line interface can be insecure.

Welcome to the MySQL monitor. Commands end with ; or \g.
Your MySQL connection id is 31
Server version: 5.7.20 MySQL Community Server (GPL)

Copyright (c) 2000, 2017, Oracle and/or its affiliates. All rights reserved.

Oracle is a registered trademark of Oracle Corporation and/or its
affiliates. Other names may be trademarks of their respective
owners.

Type 'help;' or '\h' for help. Type '\c' to clear the current input statement.

mysql> USE shop;
Database changed
#查询user表里的数据
mysql> SELECT * FROM user;
+----+-------+-----+-----+--------+-------+-----------+--------+
| id | name | sex | age | password | phone | loginName | remark |
+----+-------+-----+-----+--------+-------+-----------+--------+
3	david	女	28	111111	NULL	david	baann
4	小红	女	27	123456	NULL	xiaohong	black
5	小明	男	10	123456	NULL	xiaoming	berry
6	小刚	男	12	123456	NULL	xiaogang	banner
7	小王	男	14	111111	NULL	xiaowang	banana
8	小绿	女	34	222222	NULL	xiaolv	car
9	晓峰	男	15	333333	NULL	xiaofeng	carray
10	小影	女	26	444444	NULL	xiaoying	baaaa
11	大梅	女	27	555555	NULL	damei	accc
+----+-------+-----+-----+--------+-------+-----------+--------+
9 rows in set (0.00 sec)

#计算年龄这一列的平均年龄
mysql> SELECT avg(age) FROM user;
+----------+
| avg(age) |
+----------+
| 21.4444 |
+----------+
1 row in set (0.00 sec)

mysql>

4.4.2 COUNT() 函数

MySQL数据库中的COUNT()函数用来计算表中记录的个数或者列中值的个数，计算内容由SELECT语句指定，如要获取user表中age > 20 的个数。

SELECT count(*) FROM user WHERE age >20;

实战演练 ——COUNT()函数

Microsoft Windows [版本 6.1.7601]
版权所有 (c) 2009 Microsoft Corporation。保留所有权利。

C:\Users\Administrator>mysql –u root –p123456
mysql: [Warning] Using a password on the command line interface can be insecure.

Welcome to the MySQL monitor. Commands end with ; or \g.
Your MySQL connection id is 31
Server version: 5.7.20 MySQL Community Server (GPL)

Copyright (c) 2000, 2017, Oracle and/or its affiliates. All rights reserved.

Oracle is a registered trademark of Oracle Corporation and/or its
affiliates. Other names may be trademarks of their respective
owners.

Type 'help;' or '\h' for help. Type '\c' to clear the current input statement.

mysql> USE shop;
Database changed

mysql> SELECT * FROM user;

+----+------+-----+-----+----------+-------+-----------+--------+
| id | name | sex | age | password | phone | loginName | remark |
+----+------+-----+-----+----------+-------+-----------+--------+
3	david	女	28	111111	NULL	david	baann
4	小红	女	27	123456	NULL	xiaohong	black
5	小明	男	10	123456	NULL	xiaoming	berry
6	小刚	男	12	123456	NULL	xiaogang	banner
7	小王	男	14	111111	NULL	xiaowang	banana
8	小绿	女	34	222222	NULL	xiaolv	car
9	晓峰	男	23	333333	NULL	xiaofeng	carray
10	小影	女	26	444444	NULL	xiaoying	baaaa

```
| 11 | 大梅  | 女 | 27 | 555555 | NULL | damei    | accc |
+--+----+---+---+------+---+------+----+
9 rows in set (0.00 sec)

#计算年龄大于20的个数
mysql> SELECT count(*) FROM user WHERE age > 20;
+------+
| count(*) |
+------+
|    5 |
+------+
1 row in set (0.00 sec)

mysql>
```

4.4.3 MAX()/MIN() 函数

MySQL数据库中的MAX()函数用于选取数据中的最大值，MIN()函数用于选取数据中的最小值，如可以获取年龄的最大值或者最小值。

SELECT max(age) FROM user；

SELECT min(age) FROM user；

实战演练——MAX()/MIN() 函数

```
Microsoft Windows [版本 6.1.7601]
版权所有 (c) 2009 Microsoft Corporation。保留所有权利。

C:\Users\Administrator>mysql –u root –p123456
mysql: [Warning] Using a password on the command line interface can be insecure.

Welcome to the MySQL monitor.  Commands end with ; or \g.
Your MySQL connection id is 31
Server version: 5.7.20 MySQL Community Server (GPL)

Copyright (c) 2000, 2017, Oracle and/or its affiliates. All rights reserved.

Oracle is a registered trademark of Oracle Corporation and/or its
affiliates. Other names may be trademarks of their respective
owners.

Type 'help;' or '\h' for help. Type '\c' to clear the current input statement.

mysql> USE shop;
Database changed

mysql> SELECT * FROM user;
+--+----+---+---+------+---+------+----+
| id | name | sex | age | password | phone | loginName | remark |
+--+----+---+---+------+---+------+----+
| 3 | david | 女 | 28 | 111111 | NULL | david    | baann |
| 4 | 小红  | 女 | 27 | 123456 | NULL | xiaohong | black |
```

```
|  5 | 小明  | 男  | 10 | 123456 | NULL | xiaoming | berry  |
|  6 | 小刚  | 男  | 12 | 123456 | NULL | xiaogang | banner |
|  7 | 小王  | 男  | 14 | 111111 | NULL | xiaowang | banana |
|  8 | 小绿  | 女  | 34 | 222222 | NULL | xiaolv   | car    |
|  9 | 晓峰  | 男  | 15 | 333333 | NULL | xiaofeng | carray |
| 10 | 小影  | 女  | 26 | 444444 | NULL | xiaoying | baaaa  |
| 11 | 大梅  | 女  | 27 | 555555 | NULL | damei    | accc   |
+----+-------+-----+----+--------+------+----------+--------+
9 rows in set (0.00 sec)
```

#所有数据中的最大年龄
```
mysql> SELECT max(age) FROM user;
+---------+
| max(age)|
+---------+
|      34 |
+---------+
1 row in set (0.00 sec)
```

#男生中的最大年龄
```
mysql> SELECT max(age) FROM user WHERE sex='男';
+---------+
| max(age)|
+---------+
|      15 |
+---------+
1 row in set (0.00 sec)
```

#所有数据中的最小年龄
```
mysql> SELECT min(age) FROM user;
+---------+
| min(age)|
+---------+
|      10 |
+---------+
1 row in set (0.00 sec)
```

#女生中的最小年龄
```
mysql> SELECT min(age) FROM user WHERE sex='女';
+---------+
| min(age)|
+---------+
|      26 |
+---------+
1 row in set (0.00 sec)

mysql>
```

4.4.4 SUM()函数

　　MySQL数据库中的SUM()函数用来计算满足条件的某一列的总和，如可以计算年龄的总和，或者可以计算男生年龄的总和。聚合函数都可以设置别名。

SELECT sum(age) FROM user ;

实战演练——SUM()函数

Microsoft Windows [版本 6.1.7601]
版权所有 (c) 2009 Microsoft Corporation。保留所有权利。

C:\Users\Administrator>mysql –u root –p123456
mysql: [Warning] Using a password on the command line interface can be insecure.

Welcome to the MySQL monitor. Commands end with ; or \g.
Your MySQL connection id is 31
Server version: 5.7.20 MySQL Community Server (GPL)

Copyright (c) 2000, 2017, Oracle and/or its affiliates. All rights reserved.

Oracle is a registered trademark of Oracle Corporation and/or its
affiliates. Other names may be trademarks of their respective
owners.

Type 'help;' or '\h' for help. Type '\c' to clear the current input statement.

mysql> USE shop;
Database changed

mysql> SELECT * FROM user;

```
+----+------+------+------+----------+-------+-----------+--------+
| id | name | sex  | age  | password | phone | loginName | remark |
+----+------+------+------+----------+-------+-----------+--------+
|  3 | david| 女   |  28  | 111111   | NULL  | david     | baann  |
|  4 | 小红 | 女   |  27  | 123456   | NULL  | xiaohong  | black  |
|  5 | 小明 | 男   |  10  | 123456   | NULL  | xiaoming  | berry  |
|  6 | 小刚 | 男   |  12  | 123456   | NULL  | xiaogang  | banner |
|  7 | 小王 | 男   |  14  | 111111   | NULL  | xiaowang  | banana |
|  8 | 小绿 | 女   |  34  | 222222   | NULL  | xiaolv    | car    |
|  9 | 晓峰 | 男   |  15  | 333333   | NULL  | xiaofeng  | carray |
| 10 | 小影 | 女   |  26  | 444444   | NULL  | xiaoying  | baaaa  |
| 11 | 大梅 | 女   |  27  | 555555   | NULL  | damei     | accc   |
+----+------+------+------+----------+-------+-----------+--------+
```
9 rows in set (0.00 sec)

#年龄求和
mysql> SELECT sum(age) FROM user;
```
+----------+
| sum(age) |
+----------+
|      193 |
+----------+
```
1 row in set (0.00 sec)

#年龄求和，设置别名
mysql> SELECT sum(age) total FROM user;

```
+-----+
| total |
+-----+
|  193 |
+-----+
1 row in set (0.00 sec)

#男生求和
mysql> SELECT sum(age) manTotal FROM user WHERE sex='男';
+---------+
| manTotal |
+---------+
|      51 |
+---------+
1 row in set (0.00 sec)

mysql>
```

4.5 高级查询

MySQL提供了几种高级查询语句，用于复杂场景下进行多表一起查询，这样就会用到内连接查询、外连接查询、自然连接查询、交叉连接查询和联合查询，以满足日常业务查询的需求，从而更能体现MySQL强大的功能。

精讲视频

高级查询

4.5.1 内连接查询

内连接查询分为等值连接查询和非等值连接查询，它使用关键字inner JOIN ON，inner关键字可以省略。它是从左表中取出每一条记录，去与右表中所有的记录进行匹配，匹配成功后才会保留结果，否则不保留结果。

SELECT * FROM 左表 [INNER] JOIN 右表 ON 左表.字段 = 右表.字段；

（1）ON 关键字表示连接条件，条件字段代表相同的业务含义，如user.id和score.id。

（2）内连接可以没有连接条件，没有ON之后的内容，这样会保留所有结果。

（3）内连接可以使用WHERE 代替 ON，通常不用WHERE，因为WHERE没有ON效率高。ON指匹配到第一条成功记录后就结束，其他记录不匹配，若没有，则不进行匹配；而WHERE会一直匹配，并进行判断。

（4）通常使用字段别名、表别名，在查询数据的时候，不同表有同名字段，需要使用别名加以区分。

（5）等值连接：指使用等号"="比较两个表的连接列的值，取两表连接列的值相等的记录。

（6）非等值连接：指使用大于号">"或小于号"<"比较两个表的连接列的值，取一个表大于或小于另一个表的连接列值的记录。

使用用户表user和创建一个用户成绩表score，将用户表user的id和score表的id作为连接条件，它们相等即得到这个用户的成绩。

实战演练——内连接查询

```
Microsoft Windows [版本 6.1.7601]
版权所有 (c) 2009 Microsoft Corporation。保留所有权利。

C:\Users\Administrator>mysql –u root –p123456
mysql: [Warning] Using a password on the command line interface can be insecure.
```

```
Welcome to the MySQL monitor.  Commands end with ; or \g.
Your MySQL connection id is 156
Server version: 5.7.20 MySQL Community Server (GPL)

Copyright (c) 2000, 2017, Oracle and/or its affiliates. All rights reserved.

Oracle is a registered trademark of Oracle Corporation and/or its
affiliates. Other names may be trademarks of their respective
owners.

Type 'help;' or '\h' for help. Type '\c' to clear the current input statement.
```

mysql> USE shop;
Database changed

mysql> SELECT * FROM user;
```
+----+-------+-----+-----+----------+-------+-----------+--------+
| id | name  | sex | age | password | phone | loginName | remark |
+----+-------+-----+-----+----------+-------+-----------+--------+
|  3 | david | 女  |  28 | 111111   | NULL  | david     | baann  |
|  4 | 小红  | 女  |  27 | 123456   | NULL  | xiaohong  | black  |
|  5 | 小明  | 男  |  10 | 123456   | NULL  | xiaoming  | berry  |
|  6 | 小刚  | 男  |  12 | 123456   | NULL  | xiaogang  | banner |
|  7 | 小王  | 男  |  14 | 111111   | NULL  | xiaowang  | banana |
|  8 | 小绿  | 女  |  34 | 222222   | NULL  | xiaolv    | car    |
|  9 | 晓峰  | 男  |  15 | 333333   | NULL  | xiaofeng  | carray |
| 10 | 小影  | 女  |  26 | 444444   | NULL  | xiaoying  | baaaa  |
| 11 | 大梅  | 女  |  27 | 555555   | NULL  | damei     | accc   |
+----+-------+-----+-----+----------+-------+-----------+--------+
9 rows in set (0.08 sec)
```

#创建用户成绩表
mysql> CREATE TABLE score (
```
    -> id int not null,
    -> score varchar(255),
    -> grade varchar(255),
    -> primary key(id));
Query OK, 0 rows affected (1.83 sec)
```

#插入用户id=3的成绩
mysql> INSERT INTO score VALUES(3,'90','优秀');
Query OK, 1 row affected (0.22 sec)

#插入用户id=4的成绩
mysql> INSERT INTO score VALUES(4,'87','中等');
Query OK, 1 row affected (0.00 sec)

#插入用户id=5的成绩
mysql> INSERT INTO score VALUES(5,'70','中等');
Query OK, 1 row affected (0.00 sec)

#插入用户id=11的成绩
mysql> INSERT INTO score VALUES(11,'40','不及格');
Query OK, 1 row affected (0.05 sec)

#成绩表数据
mysql> SELECT * FROM score;

```
+----+-------+--------+
| id | score | grade  |
+----+-------+--------+
|  3 |  90   | 优秀    |
|  4 |  87   | 中等    |
|  5 |  70   | 中等    |
| 11 |  40   | 不及格  |
+----+-------+--------+
```
4 rows in set (0.00 sec)

#使用等值内连接查询
mysql> SELECT * FROM user u INNER JOIN score s ON u.id = s.id;

```
+----+------+-----+-----+----------+-------+-----------+--------+----+-------+--------+
| id | name | sex | age | password | phone | loginName | remark | id | score | grade  |
+----+------+-----+-----+----------+-------+-----------+--------+----+-------+--------+
|  3 |david | 女  | 28  | 111111   | NULL  | david     | baann  |  3 |  90   | 优秀    |
|  4 |小红  | 女  | 27  | 123456   | NULL  | xiaohong  | black  |  4 |  87   | 中等    |
|  5 |小明  | 男  | 10  | 123456   | NULL  | xiaoming  | berry  |  5 |  70   | 中等    |
| 11 |大梅  | 女  | 27  | 555555   | NULL  | damei     | accc   | 11 |  40   | 不及格  |
+----+------+-----+-----+----------+-------+-----------+--------+----+-------+--------+
```
4 rows in set (0.00 sec)

#省略关键字inner
mysql> SELECT * FROM user u JOIN score s ON u.id = s.id;

```
+----+------+-----+-----+----------+-------+-----------+--------+----+-------+--------+
| id | name | sex | age | password | phone | loginName | remark | id | score | grade  |
+----+------+-----+-----+----------+-------+-----------+--------+----+-------+--------+
|  3 |david | 女  | 28  | 111111   | NULL  | david     | baann  |  3 |  90   | 优秀    |
|  4 |小红  | 女  | 27  | 123456   | NULL  | xiaohong  | black  |  4 |  87   | 中等    |
|  5 |小明  | 男  | 10  | 123456   | NULL  | xiaoming  | berry  |  5 |  70   | 中等    |
| 11 |大梅  | 女  | 27  | 555555   | NULL  | damei     | accc   | 11 |  40   | 不及格  |
+----+------+-----+-----+----------+-------+-----------+--------+----+-------+--------+
```
4 rows in set (0.00 sec)

#使用别名，查询指定字段的数据
mysql> SELECT u.id uid,u.name,u.sex,u.age,s.score,s.grade FROM user u JOIN score s ON u.id = s.id;

```
+-----+------+-----+-----+-------+--------+
| uid | name | sex | age | score | grade  |
+-----+------+-----+-----+-------+--------+
|  3  |david | 女  | 28  |  90   | 优秀    |
|  4  |小红  | 女  | 27  |  87   | 中等    |
|  5  |小明  | 男  | 10  |  70   | 中等    |
| 11  |大梅  | 女  | 27  |  40   | 不及格  |
+-----+------+-----+-----+-------+--------+
```
4 rows in set (0.05 sec)

#省略on关键字查询
```
mysql> SELECT * FROM user u JOIN score s;
+----+--------+------+------+----------+-------+-----------+--------+----+-------+--------+
| id | name   | sex  | age  | password | phone | loginName | remark | id | score | grade  |
+----+--------+------+------+----------+-------+-----------+--------+----+-------+--------+
|  3 | david  | 女   |   28 | 111111   | NULL  | david     | baann  |  3 |    90 | 优秀   |
|  3 | david  | 女   |   28 | 111111   | NULL  | david     | baann  |  4 |    87 | 中等   |
|  3 | david  | 女   |   28 | 111111   | NULL  | david     | baann  |  5 |    70 | 中等   |
|  3 | david  | 女   |   28 | 111111   | NULL  | david     | baann  | 11 |    40 | 不及格 |
|  4 | 小红   | 女   |   27 | 123456   | NULL  | xiaohong  | black  |  3 |    90 | 优秀   |
|  4 | 小红   | 女   |   27 | 123456   | NULL  | xiaohong  | black  |  4 |    87 | 中等   |
|  4 | 小红   | 女   |   27 | 123456   | NULL  | xiaohong  | black  |  5 |    70 | 中等   |
|  4 | 小红   | 女   |   27 | 123456   | NULL  | xiaohong  | black  | 11 |    40 | 不及格 |
|  5 | 小明   | 男   |   10 | 123456   | NULL  | xiaoming  | berry  |  3 |    90 | 优秀   |
|  5 | 小明   | 男   |   10 | 123456   | NULL  | xiaoming  | berry  |  4 |    87 | 中等   |
|  5 | 小明   | 男   |   10 | 123456   | NULL  | xiaoming  | berry  |  5 |    70 | 中等   |
|  5 | 小明   | 男   |   10 | 123456   | NULL  | xiaoming  | berry  | 11 |    40 | 不及格 |
|  6 | 小刚   | 男   |   12 | 123456   | NULL  | xiaogang  | banner |  3 |    90 | 优秀   |
|  6 | 小刚   | 男   |   12 | 123456   | NULL  | xiaogang  | banner |  4 |    87 | 中等   |
|  6 | 小刚   | 男   |   12 | 123456   | NULL  | xiaogang  | banner |  5 |    70 | 中等   |
|  6 | 小刚   | 男   |   12 | 123456   | NULL  | xiaogang  | banner | 11 |    40 | 不及格 |
|  7 | 小王   | 男   |   14 | 111111   | NULL  | xiaowang  | banana |  3 |    90 | 优秀   |
|  7 | 小王   | 男   |   14 | 111111   | NULL  | xiaowang  | banana |  4 |    87 | 中等   |
|  7 | 小王   | 男   |   14 | 111111   | NULL  | xiaowang  | banana |  5 |    70 | 中等   |
|  7 | 小王   | 男   |   14 | 111111   | NULL  | xiaowang  | banana | 11 |    40 | 不及格 |
|  8 | 小绿   | 女   |   34 | 222222   | NULL  | xiaolv    | car    |  3 |    90 | 优秀   |
|  8 | 小绿   | 女   |   34 | 222222   | NULL  | xiaolv    | car    |  4 |    87 | 中等   |
|  8 | 小绿   | 女   |   34 | 222222   | NULL  | xiaolv    | car    |  5 |    70 | 中等   |
|  8 | 小绿   | 女   |   34 | 222222   | NULL  | xiaolv    | car    | 11 |    40 | 不及格 |
|  9 | 晓峰   | 男   |   15 | 333333   | NULL  | xiaofeng  | carray |  3 |    90 | 优秀   |
|  9 | 晓峰   | 男   |   15 | 333333   | NULL  | xiaofeng  | carray |  4 |    87 | 中等   |
|  9 | 晓峰   | 男   |   15 | 333333   | NULL  | xiaofeng  | carray |  5 |    70 | 中等   |
|  9 | 晓峰   | 男   |   15 | 333333   | NULL  | xiaofeng  | carray | 11 |    40 | 不及格 |
| 10 | 小影   | 女   |   26 | 444444   | NULL  | xiaoying  | baaaa  |  3 |    90 | 优秀   |
| 10 | 小影   | 女   |   26 | 444444   | NULL  | xiaoying  | baaaa  |  4 |    87 | 中等   |
| 10 | 小影   | 女   |   26 | 444444   | NULL  | xiaoying  | baaaa  |  5 |    70 | 中等   |
| 10 | 小影   | 女   |   26 | 444444   | NULL  | xiaoying  | baaaa  | 11 |    40 | 不及格 |
| 11 | 大梅   | 女   |   27 | 555555   | NULL  | damei     | accc   |  3 |    90 | 优秀   |
| 11 | 大梅   | 女   |   27 | 555555   | NULL  | damei     | accc   |  4 |    87 | 中等   |
| 11 | 大梅   | 女   |   27 | 555555   | NULL  | damei     | accc   |  5 |    70 | 中等   |
| 11 | 大梅   | 女   |   27 | 555555   | NULL  | damei     | accc   | 11 |    40 | 不及格 |
+----+--------+------+------+----------+-------+-----------+--------+----+-------+--------+
36 rows in set (0.00 sec)
```

#使用非等值内连接查询，查询u.id > s.id的数据
```
mysql> SELECT * FROM user u INNER JOIN score s ON u.id > s.id;
+----+--------+------+------+----------+-------+-----------+--------+----+-------+-------+
| id | name   | sex  | age  | password | phone | loginName | remark | id | score | grade |
+----+--------+------+------+----------+-------+-----------+--------+----+-------+-------+
|  4 | 小红   | 女   |   27 | 123456   | NULL  | xiaohong  | black  |  3 |    90 | 优秀  |
|  5 | 小明   | 男   |   10 | 123456   | NULL  | xiaoming  | berry  |  3 |    90 | 优秀  |
|  6 | 小刚   | 男   |   12 | 123456   | NULL  | xiaogang  | banner |  3 |    90 | 优秀  |
```

```
| 7 | 小王  | 男 | 14 | 111111 | NULL | xiaowang  | banana | 3 | 90 | 优秀 |
| 8 | 小绿  | 女 | 34 | 222222 | NULL | xiaolv    | car    | 3 | 90 | 优秀 |
| 9 | 晓峰  | 男 | 15 | 333333 | NULL | xiaofeng  | carray | 3 | 90 | 优秀 |
|10 | 小影  | 女 | 26 | 444444 | NULL | xiaoying  | baaaa  | 3 | 90 | 优秀 |
|11 | 大梅  | 女 | 27 | 555555 | NULL | damei     | accc   | 3 | 90 | 优秀 |
| 5 | 小明  | 男 | 10 | 123456 | NULL | xiaoming  | berry  | 4 | 87 | 中等 |
| 6 | 小刚  | 男 | 12 | 123456 | NULL | xiaogang  | banner | 4 | 87 | 中等 |
| 7 | 小王  | 男 | 14 | 111111 | NULL | xiaowang  | banana | 4 | 87 | 中等 |
| 8 | 小绿  | 女 | 34 | 222222 | NULL | xiaolv    | car    | 4 | 87 | 中等 |
| 9 | 晓峰  | 男 | 15 | 333333 | NULL | xiaofeng  | carray | 4 | 87 | 中等 |
|10 | 小影  | 女 | 26 | 444444 | NULL | xiaoying  | baaaa  | 4 | 87 | 中等 |
|11 | 大梅  | 女 | 27 | 555555 | NULL | damei     | accc   | 4 | 87 | 中等 |
| 6 | 小刚  | 男 | 12 | 123456 | NULL | xiaogang  | banner | 5 | 70 | 中等 |
| 7 | 小王  | 男 | 14 | 111111 | NULL | xiaowang  | banana | 5 | 70 | 中等 |
| 8 | 小绿  | 女 | 34 | 222222 | NULL | xiaolv    | car    | 5 | 70 | 中等 |
| 9 | 晓峰  | 男 | 15 | 333333 | NULL | xiaofeng  | carray | 5 | 70 | 中等 |
|10 | 小影  | 女 | 26 | 444444 | NULL | xiaoying  | baaaa  | 5 | 70 | 中等 |
|11 | 大梅  | 女 | 27 | 555555 | NULL | damei     | accc   | 5 | 70 | 中等 |
+---+------+---+----+--------+------+-----------+--------+---+----+------+
21 rows in set (0.00 sec)

mysql>
```

4.5.2 外连接查询

外连接查询分为左外连接查询和右外连接查询。外连接查询是以一张表为基础，取出里面的所有记录，然后将每条记录与另外一张表进行连接，不管能不能匹配上，最终都会保留。也就是说，能匹配，正确保留；不能匹配，其他表的字段都置空null，它是使用关键字outer JOIN ON进行连接查询的。

SELECT * FROM 左表 LEFT/RIGHT JOIN 右表 ON 左表.字段 = 右表.字段;

（1）LEFT JOIN 也可写成LEFT OUTER JOIN，左连接查询返回左表中符合条件的所有行，如果左表中的某行记录在右表中没有匹配，那么在结果集中，相关右表的列显示为空null。

（2）RIGHT JOIN 也可写成RIGHT OUTER JOIN，右连接查询返回所有符合条件的右表记录，如果右表中的记录在左表中找不到匹配，那么相关左表的列显示为空null。

（3）ON后面的条件内容不可以省略。

下面以user表为主和score表进行左连接和右连接。

实战演练——外连接查询

Microsoft Windows [版本 6.1.7601]
版权所有 (c) 2009 Microsoft Corporation。保留所有权利。

C:\Users\Administrator>mysql –u root –p123456
mysql: [Warning] Using a password on the command line interface can be insecure.

Welcome to the MySQL monitor. Commands end with ; or \g.
Your MySQL connection id is 157
Server version: 5.7.20 MySQL Community Server (GPL)

CopyRIGHT (c) 2000, 2017, Oracle and/or its affiliates. All RIGHTs reserved.

Oracle is a registered trademark of Oracle Corporation and/or its

affiliates. Other names may be trademarks of their respective owners.

Type 'help;' or '\h' for help. Type '\c' to clear the current input statement.

```
mysql> USE shop;
Database changed

mysql> SELECT * FROM user;
+----+-------+-----+-----+----------+-------+-----------+--------+
| id | name  | sex | age | password | phone | loginName | remark |
+----+-------+-----+-----+----------+-------+-----------+--------+
|  3 | david | 女  |  28 | 111111   | NULL  | david     | baann  |
|  4 | 小红  | 女  |  27 | 123456   | NULL  | xiaohong  | black  |
|  5 | 小明  | 男  |  10 | 123456   | NULL  | xiaoming  | berry  |
|  6 | 小刚  | 男  |  12 | 123456   | NULL  | xiaogang  | banner |
|  7 | 小王  | 男  |  14 | 111111   | NULL  | xiaowang  | banana |
|  8 | 小绿  | 女  |  34 | 222222   | NULL  | xiaolv    | car    |
|  9 | 晓峰  | 男  |  15 | 333333   | NULL  | xiaofeng  | carray |
| 10 | 小影  | 女  |  26 | 444444   | NULL  | xiaoying  | baaaa  |
| 11 | 大梅  | 女  |  27 | 555555   | NULL  | damei     | accc   |
+----+-------+-----+-----+----------+-------+-----------+--------+
9 rows in set (0.10 sec)

mysql> SELECT * FROM score;
+----+-------+-------+
| id | score | grade |
+----+-------+-------+
|  3 |    90 | 优秀  |
|  4 |    87 | 中等  |
|  5 |    70 | 中等  |
| 11 |    40 | 不及格|
+----+-------+-------+
4 rows in set (0.00 sec)

#左外连接,省略outer
mysql> SELECT * FROM user u LEFT JOIN score s ON u.id = s.id;
+----+-------+-----+-----+----------+-------+-----------+--------+------+-------+-------+
| id | name  | sex | age | password | phone | loginName | remark | id   | score | grade |
+----+-------+-----+-----+----------+-------+-----------+--------+------+-------+-------+
|  3 | david | 女  |  28 | 111111   | NULL  | david     | baann  |    3 |    90 | 优秀  |
|  4 | 小红  | 女  |  27 | 123456   | NULL  | xiaohong  | black  |    4 |    87 | 中等  |
|  5 | 小明  | 男  |  10 | 123456   | NULL  | xiaoming  | berry  |    5 |    70 | 中等  |
| 11 | 大梅  | 女  |  27 | 555555   | NULL  | damei     | accc   |   11 |    40 | 不及格|
|  6 | 小刚  | 男  |  12 | 123456   | NULL  | xiaogang  | banner | NULL | NULL  | NULL  |
|  7 | 小王  | 男  |  14 | 111111   | NULL  | xiaowang  | banana | NULL | NULL  | NULL  |
|  8 | 小绿  | 女  |  34 | 222222   | NULL  | xiaolv    | car    | NULL | NULL  | NULL  |
|  9 | 晓峰  | 男  |  15 | 333333   | NULL  | xiaofeng  | carray | NULL | NULL  | NULL  |
| 10 | 小影  | 女  |  26 | 444444   | NULL  | xiaoying  | baaaa  | NULL | NULL  | NULL  |
+----+-------+-----+-----+----------+-------+-----------+--------+------+-------+-------+
9 rows in set (0.00 sec)
```

```
#右外连接,省略outer
mysql> SELECT * FROM user u RIGHT JOIN score s ON u.id = s.id;
+----+-------+------+------+----------+-------+-----------+--------+------+-------+--------+
| id | name  | sex  | age  | password | phone | loginName | remark | id   | score | grade  |
+----+-------+------+------+----------+-------+-----------+--------+------+-------+--------+
|  3 | david | 女   |   28 | 111111   | NULL  | david     | baann  |    3 |    90 | 优秀   |
|  4 | 小红  | 女   |   27 | 123456   | NULL  | xiaohong  | black  |    4 |    87 | 中等   |
|  5 | 小明  | 男   |   10 | 123456   | NULL  | xiaoming  | berry  |    5 |    70 | 中等   |
| 11 | 大梅  | 女   |   27 | 555555   | NULL  | damei     | accc   |   11 |    40 | 不及格 |
+----+-------+------+------+----------+-------+-----------+--------+------+-------+--------+
4 rows in set (0.00 sec)

#左外连接,不省略outer
mysql> SELECT * FROM user u LEFT OUTER JOIN score s ON u.id = s.id;
+----+-------+------+------+----------+-------+-----------+--------+------+-------+--------+
| id | name  | sex  | age  | password | phone | loginName | remark | id   | score | grade  |
+----+-------+------+------+----------+-------+-----------+--------+------+-------+--------+
|  3 | david | 女   |   28 | 111111   | NULL  | david     | baann  |    3 |    90 | 优秀   |
|  4 | 小红  | 女   |   27 | 123456   | NULL  | xiaohong  | black  |    4 |    87 | 中等   |
|  5 | 小明  | 男   |   10 | 123456   | NULL  | xiaoming  | berry  |    5 |    70 | 中等   |
| 11 | 大梅  | 女   |   27 | 555555   | NULL  | damei     | accc   |   11 |    40 | 不及格 |
|  6 | 小刚  | 男   |   12 | 123456   | NULL  | xiaogang  | banner | NULL | NULL  | NULL   |
|  7 | 小王  | 男   |   14 | 111111   | NULL  | xiaowang  | banana | NULL | NULL  | NULL   |
|  8 | 小绿  | 女   |   34 | 222222   | NULL  | xiaolv    | car    | NULL | NULL  | NULL   |
|  9 | 晓峰  | 男   |   15 | 333333   | NULL  | xiaofeng  | carray | NULL | NUL   | NULL   |
| 10 | 小影  | 女   |   26 | 444444   | NULL  | xiaoying  | baaaa  | NULL | NULL  | NULL   |
+----+-------+------+------+----------+-------+-----------+--------+------+-------+--------+
9 rows in set (0.00 sec)

#右外连接,不省略outer
mysql> SELECT * FROM user u RIGHT OUTER JOIN score s ON u.id = s.id;
+----+-------+------+------+----------+-------+-----------+--------+------+-------+--------+
| id | name  | sex  | age  | password | phone | loginName | remark | id   | score | grade  |
+----+-------+------+------+----------+-------+-----------+--------+------+-------+--------+
|  3 | david | 女   |   28 | 111111   | NULL  | david     | baann  |    3 |    90 | 优秀   |
|  4 | 小红  | 女   |   27 | 123456   | NULL  | xiaohong  | black  |    4 |    87 | 中等   |
|  5 | 小明  | 男   |   10 | 123456   | NULL  | xiaoming  | berry  |    5 |    70 | 中等   |
| 11 | 大梅  | 女   |   27 | 555555   | NULL  | damei     | accc   |   11 |    40 | 不及格 |
+----+-------+------+------+----------+-------+-----------+--------+------+-------+--------+
4 rows in set (0.00 sec)

mysql>
```

4.5.3 自然连接查询

自然连接查询在连接的两个表中的列名称相等时才能使用,MySQL以同名字段作为字段进行匹配,它使用关键字natural JOIN进行自然连接,这种连接很少使用。自然连接又可以分为自然内连接和自然外连接。

SELECT * FROM左表 LEFT/ RIGHT INNER JOIN 右表 ON using(字段名)
自然内连接示例:
SELECT * FROM user NATURAL JOIN score;
自然左外连接示例:
SELECT * FROM user NATURAL LEFT JOIN score;

自然右外连接示例：
SELECT * FROM user NATURAL RIGHT JOIN score;
外连接模拟自然左外连接示例：
SELECT * FROM user LEFT JOIN score using(id);

实战演练——自然连接查询

Microsoft Windows [版本 6.1.7601]
版权所有 (c) 2009 Microsoft Corporation。保留所有权利。

C:\Users\Administrator>mysql -u root -p123456
mysql: [Warning] Using a password on the command line interface can be insecure.

Welcome to the MySQL monitor. Commands end with ; or \g.
Your MySQL connection id is 216
Server version: 5.7.20 MySQL Community Server (GPL)

Copyright (c) 2000, 2017, Oracle and/or its affiliates. All rights reserved.

Oracle is a registered trademark of Oracle Corporation and/or its
affiliates. Other names may be trademarks of their respective
owners.

Type 'help;' or '\h' for help. Type '\c' to clear the current input statement.

mysql> USE shop;
Database changed

mysql> SELECT * FROM user;
```
+----+------+-----+-----+----------+-------+-----------+--------+
| id | name | sex | age | password | phone | loginName | remark |
+----+------+-----+-----+----------+-------+-----------+--------+
|  3 | david | 女  |  28 | 111111   | NULL  | david     | baann  |
|  4 | 小红  | 女  |  27 | 123456   | NULL  | xiaohong  | black  |
|  5 | 小明  | 男  |  10 | 123456   | NULL  | xiaoming  | berry  |
|  6 | 小刚  | 男  |  12 | 123456   | NULL  | xiaogang  | banner |
|  7 | 小王  | 男  |  14 | 111111   | NULL  | xiaowang  | banana |
|  8 | 小绿  | 女  |  34 | 222222   | NULL  | xiaolv    | car    |
|  9 | 晓峰  | 男  |  15 | 333333   | NULL  | xiaofeng  | carray |
| 10 | 小影  | 女  |  26 | 444444   | NULL  | xiaoying  | baaaa  |
| 11 | 大梅  | 女  |  27 | 555555   | NULL  | damei     | accc   |
+----+------+-----+-----+----------+-------+-----------+--------+
9 rows in set (0.00 sec)
```

mysql> SELECT * FROM score;
```
+----+-------+-------+
| id | score | grade |
+----+-------+-------+
|  3 |  90   | 优秀   |
|  4 |  87   | 中等   |
|  5 |  70   | 中等   |
| 11 |  40   | 不及格 |
```

4 rows in set (0.03 sec)

#自然内连接
```
mysql> SELECT * FROM user NATURAL JOIN score;
+----+-------+-----+-----+----------+-------+-----------+--------+-------+--------+
| id | name  | sex | age | password | phone | loginName | remark | score | grade  |
+----+-------+-----+-----+----------+-------+-----------+--------+-------+--------+
|  3 | david | 女  | 28  | 111111   | NULL  | david     | baann  | 90    | 优秀   |
|  4 | 小红  | 女  | 27  | 123456   | NULL  | xiaohong  | black  | 87    | 中等   |
|  5 | 小明  | 男  | 10  | 123456   | NULL  | xiaoming  | berry  | 70    | 中等   |
| 11 | 大梅  | 女  | 27  | 555555   | NULL  | damei     | accc   | 40    | 不及格 |
+----+-------+-----+-----+----------+-------+-----------+--------+-------+--------+
4 rows in set (0.00 sec)
```

#自然左外连接
```
mysql> SELECT * FROM user NATURAL LEFT JOIN score;
+----+-------+-----+-----+----------+-------+-----------+--------+-------+--------+
| id | name  | sex | age | password | phone | loginName | remark | score | grade  |
+----+-------+-----+-----+----------+-------+-----------+--------+-------+--------+
|  3 | david | 女  | 28  | 111111   | NULL  | david     | baann  | 90    | 优秀   |
|  4 | 小红  | 女  | 27  | 123456   | NULL  | xiaohong  | black  | 87    | 中等   |
|  5 | 小明  | 男  | 10  | 123456   | NULL  | xiaoming  | berry  | 70    | 中等   |
| 11 | 大梅  | 女  | 27  | 555555   | NULL  | damei     | accc   | 40    | 不及格 |
|  6 | 小刚  | 男  | 12  | 123456   | NULL  | xiaogang  | banner | NULL  | NULL   |
|  7 | 小王  | 男  | 14  | 111111   | NULL  | xiaowang  | banana | NULL  | NULL   |
|  8 | 小绿  | 女  | 34  | 222222   | NULL  | xiaolv    | car    | NULL  | NULL   |
|  9 | 晓峰  | 男  | 15  | 333333   | NULL  | xiaofeng  | carray | NULL  | NULL   |
| 10 | 小影  | 女  | 26  | 444444   | NULL  | xiaoying  | baaaa  | NULL  | NULL   |
+----+-------+-----+-----+----------+-------+-----------+--------+-------+--------+
9 rows in set (0.00 sec)
```

#自然右外连接
```
mysql> SELECT * FROM user NATURAL RIGHT JOIN score;
+----+-------+--------+-------+-----+-----+----------+-------+-----------+--------+
| id | score | grade  | name  | sex | age | password | phone | loginName | remark |
+----+-------+--------+-------+-----+-----+----------+-------+-----------+--------+
|  3 | 90    | 优秀   | david | 女  | 28  | 111111   | NULL  | david     | baann  |
|  4 | 87    | 中等   | 小红  | 女  | 27  | 123456   | NULL  | xiaohong  | black  |
|  5 | 70    | 中等   | 小明  | 男  | 10  | 123456   | NULL  | xiaoming  | berry  |
| 11 | 40    | 不及格 | 大梅  | 女  | 27  | 555555   | NULL  | damei     | accc   |
+----+-------+--------+-------+-----+-----+----------+-------+-----------+--------+
4 rows in set (0.00 sec)
```

#外连接模拟自然左外连接
```
mysql> SELECT * FROM user LEFT JOIN score using(id);
+----+-------+-----+-----+----------+-------+-----------+--------+-------+--------+
| id | name  | sex | age | password | phone | loginName | remark | score | grade  |
+----+-------+-----+-----+----------+-------+-----------+--------+-------+--------+
|  3 | david | 女  | 28  | 111111   | NULL  | david     | baann  | 90    | 优秀   |
|  4 | 小红  | 女  | 27  | 123456   | NULL  | xiaohong  | black  | 87    | 中等   |
|  5 | 小明  | 男  | 10  | 123456   | NULL  | xiaoming  | berry  | 70    | 中等   |
```

```
| 11 | 大梅  | 女 | 27 | 555555 | NULL | damei    | accc   | 40   | 不及格 |
|  6 | 小刚  | 男 | 12 | 123456 | NULL | xiaogang | banner | NULL | NULL   |
|  7 | 小王  | 男 | 14 | 111111 | NULL | xiaowang | banana | NULL | NULL   |
|  8 | 小绿  | 女 | 34 | 222222 | NULL | xiaolv   | car    | NULL | NULL   |
|  9 | 晓峰  | 男 | 15 | 333333 | NULL | xiaofeng | carray | NULL | NULL   |
| 10 | 小影  | 女 | 26 | 444444 | NULL | xiaoying | baaaa  | NULL | NULL   |
+----+-------+----+----+--------+------+----------+--------+------+--------+
9 rows in set (0.00 sec)

mysql>
```

4.5.4 交叉连接查询

交叉连接查询是从一张表中循环取出每一条记录,然后将每条记录都在另外一张表中进行匹配,匹配的结果都会保留,而连接本身字段会增加,其最终结果称为笛卡尔积,它是使用关键字CROSS JOIN来进行交叉连接查询的,我们平时很少会使用这个交叉连接查询。

SELECT * 左表 CROSS JOIN 右表 或 FROM 左表,右表;

笛卡尔积是将两个表变成一个表,两个表的任意行自由组合,最后将其视为一个完整的临时表来进行查询操作,如图4.1所示。

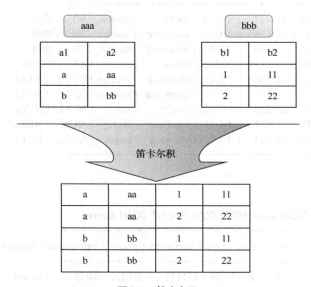

图4.1 笛卡尔积

下面将用户user表和成绩score表进行交叉连接查询。

实战演练——交叉连接查询

Microsoft Windows [版本 6.1.7601]
版权所有 (c) 2009 Microsoft Corporation。保留所有权利。

C:\Users\Administrator>mysql –u root –p123456
mysql: [Warning] Using a password on the command line interface can be insecure.

Welcome to the MySQL monitor. Commands end with ; or \g.
Your MySQL connection id is 158
Server version: 5.7.20 MySQL Community Server (GPL)

Copyright (c) 2000, 2017, Oracle and/or its affiliates. All rights reserved.

Oracle is a registered trademark of Oracle Corporation and/or its
affiliates. Other names may be trademarks of their respective
owners.

Type 'help;' or '\h' for help. Type '\c' to clear the current input statement.

mysql> USE shop;
Database changed

mysql> SELECT * FROM user;

id	name	sex	age	password	phone	loginName	remark
3	david	女	28	111111	NULL	david	baann
4	小红	女	27	123456	NULL	xiaohong	black
5	小明	男	10	123456	NULL	xiaoming	berry
6	小刚	男	12	123456	NULL	xiaogang	banner
7	小王	男	14	111111	NULL	xiaowang	banana
8	小绿	女	34	222222	NULL	xiaolv	car
9	晓峰	男	15	333333	NULL	xiaofeng	carray
10	小影	女	26	444444	NULL	xiaoying	baaaa
11	大梅	女	27	555555	NULL	damei	accc

9 rows in set (0.08 sec)

mysql> SELECT * FROM score;

id	score	grade
3	90	优秀
4	87	中等
5	70	中等
11	40	不及格

4 rows in set (0.05 sec)

#交叉连接查询，使用关键字cross JOIN
mysql> SELECT * FROM user CROSS JOIN score;

id	name	sex	age	password	phone	loginName	remark	id	score	grade
3	david	女	28	111111	NULL	david	baann	3	90	优秀
3	david	女	28	111111	NULL	david	baann	4	87	中等
3	david	女	28	111111	NULL	david	baann	5	70	中等
3	david	女	28	111111	NULL	david	baann	11	40	不及格
4	小红	女	27	123456	NULL	xiaohong	black	3	90	优秀
4	小红	女	27	123456	NULL	xiaohong	black	4	87	中等
4	小红	女	27	123456	NULL	xiaohong	black	5	70	中等

```
| 4 | 小红  | 女 | 27 | 123456 | NULL | xiaohong | black  | 11 | 40 | 不及格 |
| 5 | 小明  | 男 | 10 | 123456 | NULL | xiaoming | berry  |  3 | 90 | 优秀   |
| 5 | 小明  | 男 | 10 | 123456 | NULL | xiaoming | berry  |  4 | 87 | 中等   |
| 5 | 小明  | 男 | 10 | 123456 | NULL | xiaoming | berry  |  5 | 70 | 中等   |
| 5 | 小明  | 男 | 10 | 123456 | NULL | xiaoming | berry  | 11 | 40 | 不及格 |
| 6 | 小刚  | 男 | 12 | 123456 | NULL | xiaogang | banner |  3 | 90 | 优秀   |
| 6 | 小刚  | 男 | 12 | 123456 | NULL | xiaogang | banner |  4 | 87 | 中等   |
| 6 | 小刚  | 男 | 12 | 123456 | NULL | xiaogang | banner |  5 | 70 | 中等   |
| 6 | 小刚  | 男 | 12 | 123456 | NULL | xiaogang | banner | 11 | 40 | 不及格 |
| 7 | 小王  | 男 | 14 | 111111 | NULL | xiaowang | banana |  3 | 90 | 优秀   |
| 7 | 小王  | 男 | 14 | 111111 | NULL | xiaowang | banana |  4 | 87 | 中等   |
| 7 | 小王  | 男 | 14 | 111111 | NULL | xiaowang | banana |  5 | 70 | 中等   |
| 7 | 小王  | 男 | 14 | 111111 | NULL | xiaowang | banana | 11 | 40 | 不及格 |
| 8 | 小绿  | 女 | 34 | 222222 | NULL | xiaolv   | car    |  3 | 90 | 优秀   |
| 8 | 小绿  | 女 | 34 | 222222 | NULL | xiaolv   | car    |  4 | 87 | 中等   |
| 8 | 小绿  | 女 | 34 | 222222 | NULL | xiaolv   | car    |  5 | 70 | 中等   |
| 8 | 小绿  | 女 | 34 | 222222 | NULL | xiaolv   | car    | 11 | 40 | 不及格 |
| 9 | 晓峰  | 男 | 15 | 333333 | NULL | xiaofeng | carray |  3 | 90 | 优秀   |
| 9 | 晓峰  | 男 | 15 | 333333 | NULL | xiaofeng | carray |  4 | 87 | 中等   |
| 9 | 晓峰  | 男 | 15 | 333333 | NULL | xiaofeng | carray |  5 | 70 | 中等   |
| 9 | 晓峰  | 男 | 15 | 333333 | NULL | xiaofeng | carray | 11 | 40 | 不及格 |
|10 | 小影  | 女 | 26 | 444444 | NULL | xiaoying | baaaa  |  3 | 90 | 优秀   |
|10 | 小影  | 女 | 26 | 444444 | NULL | xiaoying | baaaa  |  4 | 87 | 中等   |
|10 | 小影  | 女 | 26 | 444444 | NULL | xiaoying | baaaa  |  5 | 70 | 中等   |
|10 | 小影  | 女 | 26 | 444444 | NULL | xiaoying | baaaa  | 11 | 40 | 不及格 |
|11 | 大梅  | 女 | 27 | 555555 | NULL | damei    | accc   |  3 | 90 | 优秀   |
|11 | 大梅  | 女 | 27 | 555555 | NULL | damei    | accc   |  4 | 87 | 中等   |
|11 | 大梅  | 女 | 27 | 555555 | NULL | damei    | accc   |  5 | 70 | 中等   |
|11 | 大梅  | 女 | 27 | 555555 | NULL | damei    | accc   | 11 | 40 | 不及格 |
+---+-------+----+----+--------+------+----------+--------+----+----+--------+
36 rows in set (0.00 sec)
```

#交叉连接查询，FROM两个表
```
mysql> SELECT * FROM user,score;
+----+-------+-----+-----+----------+-------+----------+--------+----+-------+--------+
| id | name  | sex | age | password | phone | loginName| remark | id | score | grade  |
+----+-------+-----+-----+----------+-------+----------+--------+----+-------+--------+
|  3 | david | 女  | 28  | 111111   | NULL  | david    | baann  |  3 |  90   | 优秀   |
|  3 | david | 女  | 28  | 111111   | NULL  | david    | baann  |  4 |  87   | 中等   |
|  3 | david | 女  | 28  | 111111   | NULL  | david    | baann  |  5 |  70   | 中等   |
|  3 | david | 女  | 28  | 111111   | NULL  | david    | baann  | 11 |  40   | 不及格 |
|  4 | 小红  | 女  | 27  | 123456   | NULL  | xiaohong | black  |  3 |  90   | 优秀   |
|  4 | 小红  | 女  | 27  | 123456   | NULL  | xiaohong | black  |  4 |  87   | 中等   |
|  4 | 小红  | 女  | 27  | 123456   | NULL  | xiaohong | black  |  5 |  70   | 中等   |
|  4 | 小红  | 女  | 27  | 123456   | NULL  | xiaohong | black  | 11 |  40   | 不及格 |
|  5 | 小明  | 男  | 10  | 123456   | NULL  | xiaoming | berry  |  3 |  90   | 优秀   |
|  5 | 小明  | 男  | 10  | 123456   | NULL  | xiaoming | berry  |  4 |  87   | 中等   |
|  5 | 小明  | 男  | 10  | 123456   | NULL  | xiaoming | berry  |  5 |  70   | 中等   |
|  5 | 小明  | 男  | 10  | 123456   | NULL  | xiaoming | berry  | 11 |  40   | 不及格 |
|  6 | 小刚  | 男  | 12  | 123456   | NULL  | xiaogang | banner |  3 |  90   | 优秀   |
|  6 | 小刚  | 男  | 12  | 123456   | NULL  | xiaogang | banner |  4 |  87   | 中等   |
|  6 | 小刚  | 男  | 12  | 123456   | NULL  | xiaogang | banner |  5 |  70   | 中等   |
```

```
| 6 | 小刚   | 男 | 12 | 123456 | NULL | xiaogang | banner | 11 | 40 | 不及格 |
| 7 | 小王   | 男 | 14 | 111111 | NULL | xiaowang | banana | 3  | 90 | 优秀   |
| 7 | 小王   | 男 | 14 | 111111 | NULL | xiaowang | banana | 4  | 87 | 中等   |
| 7 | 小王   | 男 | 14 | 111111 | NULL | xiaowang | banana | 5  | 70 | 中等   |
| 7 | 小王   | 男 | 14 | 111111 | NULL | xiaowang | banana | 11 | 40 | 不及格 |
| 8 | 小绿   | 女 | 34 | 222222 | NULL | xiaolv   | car    | 3  | 90 | 优秀   |
| 8 | 小绿   | 女 | 34 | 222222 | NULL | xiaolv   | car    | 4  | 87 | 中等   |
| 8 | 小绿   | 女 | 34 | 222222 | NULL | xiaolv   | car    | 5  | 70 | 中等   |
| 8 | 小绿   | 女 | 34 | 222222 | NULL | xiaolv   | car    | 11 | 40 | 不及格 |
| 9 | 晓峰   | 男 | 15 | 333333 | NULL | xiaofeng | carray | 3  | 90 | 优秀   |
| 9 | 晓峰   | 男 | 15 | 333333 | NULL | xiaofeng | carray | 4  | 87 | 中等   |
| 9 | 晓峰   | 男 | 15 | 333333 | NULL | xiaofeng | carray | 5  | 70 | 中等   |
| 9 | 晓峰   | 男 | 15 | 333333 | NULL | xiaofeng | carray | 11 | 40 | 不及格 |
|10 | 小影   | 女 | 26 | 444444 | NULL | xiaoying | baaaa  | 3  | 90 | 优秀   |
|10 | 小影   | 女 | 26 | 444444 | NULL | xiaoying | baaaa  | 4  | 87 | 中等   |
|10 | 小影   | 女 | 26 | 444444 | NULL | xiaoying | baaaa  | 5  | 70 | 中等   |
|10 | 小影   | 女 | 26 | 444444 | NULL | xiaoying | baaaa  | 11 | 40 | 不及格 |
|11 | 大梅   | 女 | 27 | 555555 | NULL | damei    | accc   | 3  | 90 | 优秀   |
|11 | 大梅   | 女 | 27 | 555555 | NULL | damei    | accc   | 4  | 87 | 中等   |
|11 | 大梅   | 女 | 27 | 555555 | NULL | damei    | accc   | 5  | 70 | 中等   |
|11 | 大梅   | 女 | 27 | 555555 | NULL | damei    | accc   | 11 | 40 | 不及格 |
+---+--------+----+----+--------+------+----------+--------+----+----+--------+
36 rows in set (0.00 sec)

mysql>
```

4.5.5 联合查询

联合查询是将多次查询结果合并起来进行拼接，其字段不会增加，要求两次查询的列数必须一致，列的类型可以不一样，多次SQL语句取出的列名可以不一致，此时以第一个SQL语句的列名为准；使用关键字UNION进行联合查询，会去掉重复的行，使用关键字UNION all进行联合查询，不会去掉重复的行；ORDEY BY不能直接使用，需要对查询语句使用括号才行，要ORDER BY生效，就必须搭配LIMIT，LIMIT使用限定的最大数即可，推荐放到所有子句之后，即对最终合并的结果来排序或筛选。

SELECT column_name FROM table1
UNION (all)
SELECT column_name FROM table2

联合查询经常用于查询同一个表，但是需求不同，如查询学生信息、男生身高升序、女生身高降序；进行多表查询，多个表的结构是完全一样的，保存的数据（结构）也是一样的。

下面查询学生表的id和成绩表的id集合，使用ORDEY BY 进行排序联合查询。

实战演练——联合查询

Microsoft Windows [版本 6.1.7601]
版权所有 (c) 2009 Microsoft Corporation。保留所有权利。

C:\Users\Administrator>mysql –u root –p123456
mysql: [Warning] Using a password on the command line interface can be insecure.

Welcome to the MySQL monitor. Commands end with ; or \g.
Your MySQL connection id is 548
Server version: 5.7.20 MySQL Community Server (GPL)

```
Copyright (c) 2000, 2017, Oracle and/or its affiliates. All rights reserved.

Oracle is a registered trademark of Oracle Corporation and/or its
affiliates. Other names may be trademarks of their respective
owners.

Type 'help;' or '\h' for help. Type '\c' to clear the current input statement.

mysql> USE shop;
Database changed

mysql> SELECT * FROM user;
+----+------+-----+-----+----------+-------+-----------+--------+
| id | name | sex | age | password | phone | loginName | remark |
+----+------+-----+-----+----------+-------+-----------+--------+
|  3 | david| 女  |  28 | 111111   | NULL  | david     | baann  |
|  4 | 小红 | 女  |  27 | 123456   | NULL  | xiaohong  | black  |
|  5 | 小明 | 男  |  10 | 123456   | NULL  | xiaoming  | berry  |
|  6 | 小刚 | 男  |  12 | 123456   | NULL  | xiaogang  | banner |
|  7 | 小王 | 男  |  14 | 111111   | NULL  | xiaowang  | banana |
|  8 | 小绿 | 女  |  34 | 222222   | NULL  | xiaolv    | car    |
|  9 | 晓峰 | 男  |  15 | 333333   | NULL  | xiaofeng  | carray |
| 10 | 小影 | 女  |  26 | 444444   | NULL  | xiaoying  | baaaa  |
| 11 | 大梅 | 女  |  27 | 555555   | NULL  | damei     | accc   |
+----+------+-----+-----+----------+-------+-----------+--------+
9 rows in set (0.11 sec)

mysql> SELECT * FROM score;
+----+-------+-------+
| id | score | grade |
+----+-------+-------+
|  3 |   90  | 优秀  |
|  4 |   87  | 中等  |
|  5 |   70  | 中等  |
| 11 |   40  | 不及格|
+----+-------+-------+
4 rows in set (0.05 sec)

#查询用户user表的id和成绩score表的id集合
mysql> SELECT id FROM user
    -> UNION
    -> SELECT id FROM score;
+----+
| id |
+----+
|  3 |
|  4 |
|  5 |
|  6 |
|  7 |
|  8 |
```

```
| 9 |
| 10 |
| 11 |
+---+
9 rows in set (0.07 sec)
```

#查询用户user表的id、name和成绩score表的id、score集合
```
mysql> SELECT id,name FROM user
    -> UNION
    -> SELECT id,score FROM score;
+----+-------+
| id | name  |
+----+-------+
| 3  | david |
| 4  | 小红  |
| 5  | 小明  |
| 6  | 小刚  |
| 7  | 小王  |
| 8  | 小绿  |
| 9  | 晓峰  |
| 10 | 小影  |
| 11 | 大梅  |
| 3  | 90    |
| 4  | 87    |
| 5  | 70    |
| 11 | 40    |
+----+-------+
13 rows in set (0.00 sec)
```

#使用ORDEY BY LIMIT需要使用括号()，否则会报错
```
mysql> SELECT id,name FROM user ORDEY BY id DESC LIMIT 0,2
    -> UNION
    -> SELECT id,score FROM score ORDEY BY id DESC LIMIT 0,2;
ERROR 1221 (HY000): Incorrect usage of UNION and ORDEY BY
```

#使用ORDEY BY LIMIT需要使用括号()
```
mysql> (SELECT id,name FROM user ORDEY BY id DESC LIMIT 0,2)
    -> UNION
    -> (SELECT id,score FROM score ORDEY BY id DESC LIMIT 0,2);
+----+-------+
| id | name  |
+----+-------+
| 11 | 大梅  |
| 10 | 小影  |
| 11 | 40    |
| 5  | 70    |
+----+-------+
4 rows in set (0.00 sec)
```

#使用ORDEY BY，推荐放到所有子句之后，即对最终合并的结果来排序或筛选
```
mysql> (SELECT id,name FROM user WHERE age >20)
    -> UNION
```

```
    -> (SELECT id,score FROM score WHERE score > 80) ORDEY BY id LIMIT 0,2;
+----+--------+
| id | name   |
+----+--------+
|  3 | david  |
|  3 | 90     |
+----+--------+
2 rows in set (0.00 sec)

mysql>
```

4.6 小结

本章是非常重要的一章，MySQL查询也是必须要掌握的内容。在实际项目过程中，查询是使用最频繁的一个操作，可以应用到各个场景下，以满足业务的需求。本章要学会基本查询过滤、条件查询过滤、模糊查询过滤、字段控制查询过滤、正则表达式查询过滤这5种数据过滤方式；学会使用子查询，掌握按返回结果分类的子查询、按对返回结果的调用方法分类的子查询的使用；学会使用聚合函数，包括AVG()查询平均数函数，COUNT查询数量函数，MAX()/MIN()查询最大值、最小值函数，SUM求和函数；掌握高级查询的使用，包括内连接查询、外连接查询、自然连接查询、交叉连接查询、联合查询，用于满足复杂场景的业务使用。

第5章
MySQL函数和存储过程

本章要点

- MySQL流程控制函数
- MySQL常用函数
- 自定义函数
- 存储过程
- 自定义函数和存储过程的区别
- 小结

■ MySQL数据库提供了函数和存储过程功能。函数可以分为流程控制函数、常用函数、自定义函数。流程控制函数可以在SQL语句中实现条件选择，提高执行效率。MySQL数据库内置了一些常用函数，包括数学函数、字符串函数、日期和时间函数、系统信息函数、加密函数、格式化函数。这些函数给我们日常开发和数据操作带来了很大的便利。自定义函数用来创建自己常用的一些函数，方便每次使用。存储过程是一组为了完成特定功能的SQL语句集，经编译后存储在数据库中，用户通过指定存储过程的名字并给定参数（如果该存储过程带有参数）来调用执行它，提高数据库的执行效率。

5.1 MySQL流程控制函数

MySQL流程控制函数用于控制SQL语句中实现条件的选择，它包括以下4个函数。

MySQL流程控制函数

（1）IF函数。

IF(expr1,expr2,expr3)：如果expr1是真，返回expr2，否则返回expr3；

SELECT name,IF(age>18,'成年','未成年') FROM user；

（2）IFNULL函数。

IFNULL(expr1,expr2)：如果expr1不是NULL,返回expr1,否则返回expr2；

SELECT name,IFNULL(age,0) FROM user；

（3）NULLIF函数。

NULLIF(expr1,expr2)：如果expr1 = expr2 成立，返回值为NULL，否则返回值为expr1；

SELECT name,NULLIF(name,loginName) FROM user；

（4）CASE函数。

CASE[expr] WHEN [value] THEN[result]… ELSE[default] END：如果value是真，返回result，否则返回default；

SELECT name, CASE sex WHEN '女' THEN '女生' WHEN '男' THEN '男生' ELSE '未知' END as result FROM user；

实战演练 —— 使用流程控制函数

下面使用流程控制函数来完成。

（1）判断年龄大于18岁时，输出成年，否则输出未成年。

（2）判断年龄为NULL时，输出默认值为0。

（3）判断用户名name和登录名loginName是否相等，相等则返回NULL，否则返回用户。

（4）判断性别为女时，输出女生；判断性别为男时，输出男生。

```
Microsoft Windows [版本 6.1.7601]
版权所有 (c) 2009 Microsoft Corporation。保留所有权利。

C:\Users\Administrator>mysql -uroot -p123456
mysql: [Warning] Using a password on the command line interface can be insecure.

Welcome to the MySQL monitor.  Commands end with ; or \g.
Your MySQL connection id is 31
Server version: 5.7.20 MySQL Community Server (GPL)

Copyright (c) 2000, 2017, Oracle and/or its affiliates. All rights reserved.

Oracle is a registered trademark of Oracle Corporation and/or its
affiliates. Other names may be trademarks of their respective
owners.

Type 'help;' or '\h' for help. Type '\c' to clear the current input statement.

mysql> USE shop;
Database changed
```

```
mysql> SELECT * FROM user;
+----+--------+-----+-----+----------+-------+-----------+--------+
| id | name   | sex | age | password | phone | loginName | remark |
+----+--------+-----+-----+----------+-------+-----------+--------+
|  3 | david  | 女  | 28  | 111111   | NULL  | david     | baann  |
|  4 | 小红   | 女  | 27  | 123456   | NULL  | xiaohong  | black  |
|  5 | 小明   | 男  | 10  | 123456   | NULL  | xiaoming  | berry  |
|  6 | 小刚   | 男  | 12  | 123456   | NULL  | xiaogang  | banner |
|  7 | 小王   | 男  | 14  | 111111   | NULL  | xiaowang  | banana |
|  8 | 小绿   | 女  | 34  | 222222   | NULL  | xiaolv    | car    |
|  9 | 晓峰   | 男  | 15  | 333333   | NULL  | xiaofeng  | carray |
| 10 | 小影   | 女  | 26  | 444444   | NULL  | xiaoying  | baaaa  |
| 11 | 大梅   | 女  | 27  | 555555   | NULL  | damei     | accc   |
+----+--------+-----+-----+----------+-------+-----------+--------+
9 rows in set (0.49 sec)
```

#使用IF函数，判断年龄大于18岁时，输出成年，否则输出未成年
```
mysql> SELECT name,IF(age>18,'成年','未成年') FROM user;
+--------+-----------------------------+
| name   | IF(age>18,'成年','未成年')  |
+--------+-----------------------------+
| david  | 成年                        |
| 小红   | 成年                        |
| 小明   | 未成年                      |
| 小刚   | 未成年                      |
| 小王   | 未成年                      |
| 小绿   | 成年                        |
| 晓峰   | 未成年                      |
| 小影   | 成年                        |
| 大梅   | 成年                        |
+--------+-----------------------------+
9 rows in set (0.16 sec)
```

#使用IFNULL函数，判断年龄为NULL时，输出默认值为0
```
mysql> SELECT name,IFNULL(age,0) FROM user;
+--------+---------------+
| name   | IFNULL(age,0) |
+--------+---------------+
| david  |            28 |
| 小红   |            27 |
| 小明   |            10 |
| 小刚   |            12 |
| 小王   |            14 |
| 小绿   |            34 |
| 晓峰   |            15 |
| 小影   |            26 |
| 大梅   |            27 |
+--------+---------------+
9 rows in set (0.11 sec)
```

#使用NULLIF函数，判断用户名name和登录名loginName是否相等，相等则返回NULL，否则返回用户名name
```
mysql> SELECT name,NULLIF(name,loginName) FROM user;
```

```
+------+----------------------+
| name | NULLIF(name,loginName) |
+------+----------------------+
| david | NULL                |
| 小红  | 小红                 |
| 小明  | 小明                 |
| 小刚  | 小刚                 |
| 小王  | 小王                 |
| 小绿  | 小绿                 |
| 晓峰  | 晓峰                 |
| 小影  | 小影                 |
| 大梅  | 大梅                 |
+------+----------------------+
9 rows in set (0.05 sec)
```

#使用CASE函数，判断性别为女时，输出女生，判断性别为男时，输出男生

```
mysql> SELECT name, CASE sex WHEN '女' THEN '女生' WHEN '男' THEN '男生' ELSE '未知' END as result FROM user;
+------+--------+
| name | result |
+------+--------+
| david | 女生  |
| 小红  | 女生  |
| 小明  | 男生  |
| 小刚  | 男生  |
| 小王  | 男生  |
| 小绿  | 女生  |
| 晓峰  | 男生  |
| 小影  | 女生  |
| 大梅  | 女生  |
+------+--------+
9 rows in set (0.02 sec)

mysql>
```

5.2 MySQL常用函数

MySQL数据库内置了一些常用函数，包括数学函数、字符串函数、日期和时间函数、系统信息函数、加密函数和格式化函数。这些常用函数方便了我们日常对数据库、对表以及查询等的操作。

5.2.1 数学函数

数学函数主要用来处理与数字有关的函数，包括整数、浮点数等数学操作。数学函数如表5.1所示。

精讲视频

MySQL常用函数

表5.1 数学函数

函数名称	含义	示例
ABS(x)	返回x的绝对值	SELECT ABS(-12.3);
BIN(x)	返回x的二进制（OCT返回八进制，HEX返回十六进制）	SELECT BIN(5);

续表

函数名称	含义	示例
CEILING(x)	返回大于x的最小整数值	SELECT CEILING(13.6);
EXP(x)	返回值e（自然对数的底数）的x次方	SELECT EXP(2);
FLOOR(x)	返回小于x的最大整数值	SELECT FLOOR(12.7);
GREATEST(x1,x2,…,xn)	返回集合中最大的值	SELECT GREATEST(1,20,33,78);
LEAST(x1,x2,…,xn)	返回集合中最小的值	SELECT LEAST(1,20,33);
LN(x)	返回x的自然对数	SELECT LN(20);
LOG(x,y)	返回x的以y为底数的对数	SELECT LOG(2,10);
MOD(x,y)	返回x/y的模（余数）	SELECT MOD(3,5);
PI()	返回pi的值（圆周率）	SELECT PI();
RAND()	返回0到1内的随机值，可以通过提供一个参数（种子）使RAND()随机数生成器生成一个指定的值	SELECT RAND();
ROUND(x,y)	返回参数x的四舍五入的有y位小数的值	SELECT ROUND(123.678,2);
SIGN(x)	返回代表数字x的符号的值	SELECT SIGN(10);
SQRT(x)	返回一个数的平方根	SELECT SQRT(10);
TRUNCATE(x,y)	返回数字x截短为y位小数的结果	SELECT TRUNCATE(5642.356,2);

常用的数学函数如下。

（1）CEILING(x)：返回大于x的最小整数值，它是向上取整。

（2）FLOOR(x)：返回小于x的最大整数值，它是向下取整。

（3）ROUND(x,y)：返回参数x的四舍五入的有y位小数的值，进行四舍五入，保留y位小数。

（4）TRUNCATE(x,y)：返回数字x截短为y位小数的结果，不进行四舍五入，直接保留y位小数。

（5）MOD(x,y)：返回x/y的模，也是取余数，和x%y是等价的。

实战演练 ——使用数学函数

下面操作一下12.6向上取整，27.4向下取整，54 362.879 2四舍五入、保留2位小数，678.347 8不四舍五入、保留2位小数，3对5取余数。

```
Microsoft Windows [版本 6.1.7601]
版权所有 (c) 2009 Microsoft Corporation。保留所有权利。

C:\Users\Administrator>mysql -uroot -p123456
mysql: [Warning] Using a password on the command line interface can be insecure.

Welcome to the MySQL monitor.  Commands end with ; or \g.
Your MySQL connection id is 32
Server version: 5.7.20 MySQL Community Server (GPL)

Copyright (c) 2000, 2017, Oracle and/or its affiliates. All rights reserved.

Oracle is a registered trademark of Oracle Corporation and/or its
affiliates. Other names may be trademarks of their respective
```

owners.

Type 'help;' or '\h' for help. Type '\c' to clear the current input statement.

#12.6向上取整
```
mysql> SELECT CEILING(12.6);
+---------------+
| CEILING(12.6) |
+---------------+
|            13 |
+---------------+
1 row in set (0.00 sec)
```

#27.4向下取整
```
mysql> SELECT FLOOR(27.4);
+-------------+
| FLOOR(27.4) |
+-------------+
|          27 |
+-------------+
1 row in set (0.00 sec)
```

#54 362.879 2四舍五入、保留2位小数
```
mysql> SELECT ROUND(54362.8792,2);
+---------------------+
| ROUND(54362.8792,2) |
+---------------------+
|            54362.88 |
+---------------------+
1 row in set (0.00 sec)
```

#678.347 8不四舍五入、保留2位小数
```
mysql> SELECT TRUNCATE(678.3478,2);
+----------------------+
| TRUNCATE(678.3478,2) |
+----------------------+
|               678.34 |
+----------------------+
1 row in set (0.00 sec)
```

#3对5取余数
```
mysql> SELECT MOD(3,5);
+----------+
| MOD(3,5) |
+----------+
|        3 |
+----------+
1 row in set (0.00 sec)
```

#3对5取余数
```
mysql> SELECT 3%5;
+------+
```

```
| 3%5  |
+------+

#3除以5
mysql> SELECT 3/5;
+------+
| 3/5  |
+------+
| 0.6000 |
+------+
1 row in set (0.00 sec)
```

5.2.2 字符串函数

字符串函数是对字符以及字符串进行的一系列操作。MySQL提供了很多字符串函数，如表5.2所示。

表5.2 字符串函数

函数名称	含义	示例
ASCII(char)	返回字符的ASCII码值	SELECT ASCII('A');
CHAR_LENGTH(str)	返回字符串str的位数	SELECT CHAR_LENGTH('abc');
BIT_LENGTH(str)	返回字符串的比特长度	SELECT BIT_LENGTH('hello');
CONCAT(s1,s2…,sn)	将s1，s2，…，sn连接成字符串	SELECT CONCAT('I',' LOVE',' YOU');
CONCAT_WS(sep,s1,s2,…sn)	将s1，s2，…，sn连接成字符串，并用sep字符间隔	SELECT ONCAT_WS('-','hello','mysql');
INSERT(str,x,y,instr)	将字符串str从第x位置开始，y个字符长的子串替换为字符串instr，返回结果	SELECT INSERT('hello',2,3,'aaa');
FIND_IN_SET(str,list)	分析逗号分隔的list列表，如果发现str，返回str在list中的位置	SELECT FIND_IN_SET('he','a,b,he,ef,d');
LCASE(str)、LOWER(str)	返回将字符串str中的所有字符改变为小写后的结果	SELECT LOWER('ABC');
UCASE(str)、UPPER(str)	返回将字符串str中的所有字符改变为大写后的结果	SELECT UPPER('mysql');
LEFT(str,x)	返回字符串str中最左边的x个字符	SELECT LEFT('hello',2);
RIGHT(str,x)	返回字符串str中最右边的x个字符	SELECT RIGHT('hello',3);
LPAD(s1,len,s2)	字符串s2来填充s1的开始处，使字符串的长度达到len	SELECT LPAD('hello','10','a');
RPAD(s1,len,s2)	字符串s2来填充s1的结尾处，使字符串的长度达到len	SELECT RPAD('hello','10','a');
LENGTH(str)	返回字符串str中的字符数	SELECT LENGTH('hello');
POSITION(substr IN str)	返回子串substr在字符串str中第一次出现的位置	SELECT POSITION('he'IN 'love hello');
QUOTE(str)	用反斜杠转义str中的单引号	SELECT QUOTE('hello);
REPEAT(str,x)	返回字符串str重复x次的结果	SELECT REPEAT('hello',2);
REPLACE(s,s1,s2)	用字符串s2代替字符串s中的字符串s1	SELECT REPLACE ('hello', 'e', 'abc');

续表

函数名称	含义	示例
REVERSE(str)	返回颠倒字符串str的结果	SELECT REVERSE('hello');
LTRIM(str)	去掉字符串str开头的空格	SELECT LTRIM(' hello');
RTRIM(str)	去掉字符串str尾部的空格	SELECT RTRIM(' hello ');
TRIM(str)	去掉字符串首部和尾部的所有空格	SELECT TRIM(' hello ');
TRIM(s1 FROM s)	去掉字符串s开始处和结尾处的字符串s1	SELECT TRIM('$' FROM '$$hello$$$ab$$$');
SPACE(n)	返回空格n次	SELECT SPACE(10);
STRCMP(s1,s2)	比较字符串s1,s2	SELECT STRCMP('hello','hello world');
SUBSTRING(s,n,len)	获取从字符串s中第n个位置开始长度为len的字符串	SELECT SUBSTRING('hello world',2,5);
ELT(n,s1,s2,…)	返回第n个字符串	SELECT ELT(2,'A','B','C','D');
FIELD(s,s1,s2,…)	返回第一个与字符串s匹配的字符串的位置	SELECT FIELD('he','ho','hello','he');

常用的字符串函数如下。

（1）LENGTH(str)：获取字符串的长度。

（2）LOWER(str)、UPPER(str)：进行大小写字母的转换。

（3）STRCMP(s1,s2)：比较两个字符串的大小（开头字母的顺序）hello>yes:1 hello<yes:-1 等于：0。

（4）REPLACE(s,s1,s2)：替换字符串。

（5）CONCAT(s1,s2,…,sn)：合并拼接字符串。

（6）CONCAT_WS(sep,s1,s2,…,sn)：使用分隔符合并拼接字符串。

（7）LTRIM(str)、RTRIM(str)、TRIM(str)：去除空格。

（8）SUBSTRING(s,n,len)：截取字符串。

实战演练——使用字符串函数

下面完成获取字符串"hello"的长度；将"hello"字母变为大写；将"WORLD"字母变为小写；比较"hello"和"world"的大小；将字符串"hello"里的字母"e"替换为"abc"；将"hello"和"world"拼接成一个字符串；将"hello"和"world"使用分隔符"-"拼接成一个字符串；截取"helloworld"字符串，从第2个位置截取长度为5的字符串。

```
Microsoft Windows [版本 6.1.7601]
版权所有 (c) 2009 Microsoft Corporation。保留所有权利。

C:\Users\Administrator>mysql -uroot -p123456
mysql: [Warning] Using a password on the command line interface can be insecure.

Welcome to the MySQL monitor.  Commands end with ; or \g.
Your MySQL connection id is 34
Server version: 5.7.20 MySQL Community Server (GPL)

Copyright (c) 2000, 2017, Oracle and/or its affiliates. All rights reserved.

Oracle is a registered trademark of Oracle Corporation and/or its
affiliates. Other names may be trademarks of their respective
```

owners.

Type 'help;' or '\h' for help. Type '\c' to clear the current input statement.

```
#获取字符串"hello"的长度
mysql> SELECT LENGTH('hello');
+-----------------+
| LENGTH('hello') |
+-----------------+
|               5 |
+-----------------+
1 row in set (0.00 sec)

#将"hello"字母变为大写
mysql> SELECT UPPER('hello');
+----------------+
| UPPER('hello') |
+----------------+
| HELLO          |
+----------------+
1 row in set (0.00 sec)

#将"WORLD"字母变为小写
mysql> SELECT LOWER('WORLD');
+----------------+
| LOWER('WORLD') |
+----------------+
| world          |
+----------------+
1 row in set (0.00 sec)

#比较"hello"和"world"的大小
mysql> SELECT STRCMP('hello','world');
+-------------------------+
| STRCMP('hello','world') |
+-------------------------+
|                      -1 |
+-------------------------+
1 row in set (0.00 sec)

#将字符串"hello"里的字母"e"替换为"abc"
mysql> SELECT REPLACE('hello','e','abc');
+----------------------------+
| REPLACE('hello','e','abc') |
+----------------------------+
| habcllo                    |
+----------------------------+
1 row in set (0.00 sec)

#将"hello"和"world"拼接成一个字符串
mysql> SELECT CONCAT('hello','world');
+-------------------------+
```

```
| CONCAT('hello','world') |
+-------------------------+
| helloworld              |
+-------------------------+
1 row in set (0.00 sec)
```

#将"hello"和"world"使用分隔符"-"拼接成一个字符串
```
mysql> SELECT CONCAT_WS('-','hello','world');
+--------------------------------+
| CONCAT_WS('-','hello','world') |
+--------------------------------+
| hello-world                    |
+--------------------------------+
1 row in set (0.00 sec)
```

#截取"helloworld"字符串，从第2个位置截取长度为5的字符串
```
mysql> SELECT SUBSTRING('helloworld',2,5);
+-----------------------------+
| SUBSTRING('helloworld',2,5) |
+-----------------------------+
| ellow                       |
+-----------------------------+
1 row in set (0.00 sec)

mysql>
```

5.2.3 日期和时间函数

通过日期和时间函数可以获取当前日期、当前时间、年份、月份、天、小时等关于日期的函数。日期和时间函数如表5.3所示。

表5.3 日期和时间函数

函数名称	含义	示例
CURDATE()、CURRENT_DATE()	返回当前日期，如2017-11-30	SELECT CURDATE();
CURTIME()、CURRENT_TIME()	返回当前时间，如12:36:52	SELECT CURTIME();
NOW()、CURRENT_TIMESTAMP()、LOCALTIME()、SYSDATE()、LOCALTIMESTAMP()	返回当前日期和时间	SELECT NOW();
UNIX_TIMESTAMP()	以UNIX时间戳的形式返回当前时间	SELECT UNIX_TIMESTAMP();
UNIX_TIMESTAMP(d)	将时间d以UNIX时间戳的形式返回	SELECT UNIX_TIMESTAMP ('2017-11-30');
FROM_UNIXTIME(d)	把UNIX时间戳的时间转换为普通格式时间	SELECT FROM_UNIXTIME(2017);
UTC_DATE()	返回UTC日期	SELECT UTC_DATE();
UTC_TIME()	返回UTC时间	SELECT UTC_TIME();

续表

函数名称	含义	示例
MONTH(d)	返回日期d中的月份值，范围为1～12	SELECT MONTH('2017-11-22');
MONTHNAME(d)	返回日期d中的月份名称，如January	SELECT MONTHNAME('2017-11-30');
DAYNAME(d)	返回日期d是星期几，如Monday	SELECT DAYNAME('2017-11-29');
DAYOFWEEK(d)	返回日期d是星期几，如1是星期一	SELECT DAYOFWEEK ('2017-11-29');
WEEKDAY(d)	返回日期d是星期几，如0是星期一	SELECT WEEKDAY ('2017-11-29');
WEEK(d)	计算日期d是本年的第几周，范围为0～53	SELECT WEEK ('2017-11-29');
WEEKOFYEAR(d)	计算日期d是本年的第几周，范围为1～54	SELECT WEEKOFYEAR('2017-11-30');
DAYOFYEAR(d)	计算日期d是本年的第几天	SELECT DAYOFYEAR ('2017-11-30');
DAYOFMONTH(d)	计算日期d是本月的第几天	SELECT DAYOFMONTH ('2017-11-30');
YEAR(d)	返回日期d的年份值	SELECT YEAR ('2017-11-30');
QUARTER(d)	返回日期d是第几个季度，范围为1～4	SELECT QUARTER ('2017-11-30');
HOUR(t)	返回时间t中的小时值	SELECT HOUR('22:30:11');
MINUTE(t)	返回时间t中的分钟值	SELECT MINUTE ('22:30:11');
SECOND(t)	返回时间t中的秒钟值	SELECT SECOND ('22:30:11');
EXTRACT(type FROM d)	从日期d中获取指定值，type指定返回的值，如YEAR、HOUR等	SELECT EXTRACT(YEAR FROM '2017-11-20');
TIME_TO_SEC(t)	将时间t转换为秒	SELECT TIME_TO_SEC ('22:30:11');
SEC_TO_TIME(s)	将以秒为单位的时间s转换为时分秒的格式	SELECT SEC_TO_TIME ('22:30:11');
DATEDIFF(d1,d2)	计算日期d1～d2相隔的天数	SELECT DATEDIFF('2017-11- 20','2017-11-12');
DATE_ADD()	时间加	SELECT DATE_ADD(NOW(),interval 3 year);
DATE_SUB()	时间减	SELECT DATE_SUB(NOW(),interval 3 year);

常用的日期和时间函数如下。

（1）NOW()：当前日期和时间，如2017-11-29 23:21:19。

（2）CURDATE()：当前日期，如2017-11-29。

（3）CURTIME()：当前时间，如23:22:49。

（4）YEAR(d)：提取日期中的年份，如YEAR('2017-11-30')。

（5）MONTH(d)：提取日期中的月份，如MONTH('20170819')。

（6）DAYOFYEAR(d)：提取日期里一年中的第几天，如DAYOFYEAR('2017-11-30')。

（7）DAYOFWEEK(d)：提取日期里一星期中的第几天，如DAYOFWEEK ('2017-11-30')。

（8）HOUR(d)：提取时间中的小时，如HOUR('11:20')。

（9）MINUTE(d)：提取时间中的分钟，如MINUTE ('11:30')。

（10）SECOND(d)：提取时间中的秒数，如SECOND ('11:30:59')。

（11）DATE_ADD()：向后推时间。DATE_ADD(NOW(),INTERVAL 3 YEAR)表示当前时间往后推3年；DATE_ADD(NOW(),INTERVAL 3 MONTH)表示当前时间往后推3个月；DATE_ADD(NOW(),INTERVAL 3 DAY)表示当前时间往后推3天。

(12) DATE_SUB：向前推时间。DATE_SUB(NOW(),INTERVAL 3 DAY)表示当前时间往前推3天；DATE_SUB(NOW(),INTERVAL 3 MONTH)表示当前时间往前推3个月；DATE_SUB(NOW(),INTERVAL 3 YEAR)表示当前时间往前推3年。

实战演练——使用日期和时间函数

下面完成获取当前日期和时间、日期、时间、年份、月份、天、小时、分钟、秒数、向后推时间、向前推时间。

```
Microsoft Windows [版本 6.1.7601]
版权所有 (c) 2009 Microsoft Corporation。保留所有权利。

C:\Users\Administrator>mysql -uroot -p123456
mysql: [Warning] Using a password on the command line interface can be insecure.

Welcome to the MySQL monitor.  Commands end with ; or \g.
Your MySQL connection id is 35
Server version: 5.7.20 MySQL Community Server (GPL)

Copyright (c) 2000, 2017, Oracle and/or its affiliates. All rights reserved.

Oracle is a registered trademark of Oracle Corporation and/or its
affiliates. Other names may be trademarks of their respective
owners.

Type 'help;' or '\h' for help. Type '\c' to clear the current input statement.

#获取当前日期和时间
mysql> SELECT NOW();
+---------------------+
| NOW()               |
+---------------------+
| 2017-11-29 23:38:58 |
+---------------------+
1 row in set (0.00 sec)

#获取当前日期
mysql> SELECT CURDATE();
+------------+
| CURDATE()  |
+------------+
| 2017-11-29 |
+------------+
1 row in set (0.00 sec)

#获取当前时间
mysql> SELECT CURTIME();
+-----------+
| CURTIME() |
+-----------+
| 23:39:24  |
+-----------+
```

1 row in set (0.00 sec)

#获取日期中的年份
mysql> SELECT YEAR('2017-11-30');
+--------------------+
| YEAR('2017-11-30') |
+--------------------+
| 2017 |
+--------------------+
1 row in set (0.00 sec)

#获取日期中的月份
mysql> SELECT MONTH('20170819');
+-------------------+
| MONTH('20170819') |
+-------------------+
| 8 |
+-------------------+
1 row in set (0.00 sec)

#获取日期中这一年的第几天
mysql> SELECT DAYOFYEAR('2017-11-30');
+-------------------------+
| DAYOFYEAR('2017-11-30') |
+-------------------------+
| 334 |
+-------------------------+
1 row in set (0.00 sec)

#获取一周中的第几天
mysql> SELECT DAYOFWEEK ('2017-11-30');
+--------------------------+
| DAYOFWEEK ('2017-11-30') |
+--------------------------+
| 5 |
+--------------------------+
1 row in set (0.00 sec)

#获取时间中的小时
mysql> SELECT HOUR('11:20');
+---------------+
| HOUR('11:20') |
+---------------+
| 11 |
+---------------+
1 row in set (0.00 sec)

#获取时间中的分钟
mysql> SELECT MINUTE ('11:30:59');
+---------------------+
| MINUTE ('11:30:59') |
+---------------------+

```
|             30 |
+----------------+
1 row in set (0.00 sec)

#获取时间中的秒数
mysql> SELECT SECOND ('11:30:59');
+---------------------+
| SECOND ('11:30:59') |
+---------------------+
|                  59 |
+---------------------+
1 row in set (0.00 sec)

#将时间往后推3年
mysql> SELECT DATE_ADD(NOW(),INTERVAL 3 YEAR);
+---------------------------------+
| DATE_ADD(NOW(),INTERVAL 3 YEAR) |
+---------------------------------+
| 2020-11-29 23:41:45             |
+---------------------------------+
1 row in set (0.00 sec)

#将时间往后推3个月
mysql> SELECT DATE_ADD(NOW(),INTERVAL 3 MONTH);
+----------------------------------+
| DATE_ADD(NOW(),INTERVAL 3 MONTH) |
+----------------------------------+
| 2018-02-28 23:42:31              |
+----------------------------------+
1 row in set (0.00 sec)

#将时间往后推3天
mysql> SELECT DATE_ADD(NOW(),INTERVAL 3 DAY);
+--------------------------------+
| DATE_ADD(NOW(),INTERVAL 3 DAY) |
+--------------------------------+
| 2017-12-02 23:42:35            |
+--------------------------------+
1 row in set (0.00 sec)

#将时间往前推3天
mysql> SELECT DATE_SUB(NOW(),INTERVAL 3 DAY);
+--------------------------------+
| DATE_SUB(NOW(),INTERVAL 3 DAY) |
+--------------------------------+
| 2017-11-26 23:43:03            |
+--------------------------------+
1 row in set (0.00 sec)

#将时间往前推3个月
mysql> SELECT DATE_SUB(NOW(),INTERVAL 3 MONTH);
+----------------------------------+
```

```
| DATE_SUB(NOW(),INTERVAL 3 MONTH) |
+------------------------+
| 2017-08-29 23:43:59    |
+------------------------+
1 row in set (0.00 sec)
```

#将时间往前推3年
```
mysql> SELECT DATE_SUB(NOW(),INTERVAL 3 YEAR);
+-----------------------+
| DATE_SUB(NOW(),INTERVAL 3 YEAR) |
+-----------------------+
| 2014-11-29 23:44:06   |
+-----------------------+
1 row in set (0.00 sec)

mysql>
```

5.2.4 系统信息函数

通过系统信息函数可以获取数据库的版本号、服务器的连接数、当前数据库名、当前用户、字符串的编码集，如表5.4所示。

表5.4 系统信息函数

函 数 名 称	含 义	示 例
VERSION()	返回数据库的版本号	SELECT VERSION();
CONNECTION_ID()	返回服务器的连接数	SELECT CONNECTION_ID();
DATABASE()、SCHEMA()	返回当前数据库名	SELECT SCHEMA();
USER()、SYSTEM_USER()、SESSION_USER()	返回当前用户、系统用户、回话用户	SELECT USER();
CURRENT_USER()	返回当前用户	SELECT CURRENT_USER();
CHARSET(str)	返回字符串str的字符集	SELECT CHARSET('hello');
COLLACTION(str)	返回字符串str的字符排列方式	
LAST_INSERT_ID()	返回最近生成的AUTO_INCREMENET值	SELECT LAST_INSERT_ID();

实战演练——使用系统信息函数

下面完成获取数据库的版本号、服务器的连接数、当前数据库名、当前用户、字符串的编码集。

Microsoft Windows [版本 6.1.7601]
版权所有 (c) 2009 Microsoft Corporation。保留所有权利。

C:\Users\Administrator>mysql –uroot –p123456
mysql: [Warning] Using a password on the command line interface can be insecure.

Welcome to the MySQL monitor. Commands end with ; or \g.
Your MySQL connection id is 37
Server version: 5.7.20 MySQL Community Server (GPL)

Copyright (c) 2000, 2017, Oracle and/or its affiliates. All rights reserved.

```
Oracle is a registered trademark of Oracle Corporation and/or its
affiliates. Other names may be trademarks of their respective
owners.

Type 'help;' or '\h' for help. Type '\c' to clear the current input statement.

#当前数据库的版本号
mysql> SELECT VERSION();
+-------+
| VERSION() |
+-------+
| 5.7.20 |
+-------+
1 row in set (0.00 sec)

#服务器的连接数
mysql> SELECT CONNECTION_ID();
+---------+
| CONNECTION_ID() |
+---------+
|          37 |
+---------+
1 row in set (0.00 sec)

#当前数据库名
mysql> SELECT SCHEMA();
+-------+
| SCHEMA() |
+-------+
| NULL   |
+-------+
1 row in set (0.00 sec)

#使用shop数据库
mysql> USE shop;
Database changed

#当前数据库名
mysql> SELECT SCHEMA();
+-------+
| SCHEMA() |
+-------+
| shop   |
+-------+
1 row in set (0.00 sec)

#当前用户
mysql> SELECT USER();
+---------+
| USER()  |
+---------+
```

```
| root@localhost |
+----------------+
1 row in set (0.00 sec)

#当前用户
mysql> SELECT CURRENT_USER();
+----------------+
| CURRENT_USER() |
+----------------+
| root@localhost |
+----------------+
1 row in set (0.00 sec)

#字符串编码集
mysql> SELECT CHARSET('hello');
+------------------+
| CHARSET('hello') |
+------------------+
| utf8             |
+------------------+
1 row in set (0.00 sec)

#最近生成的AUTO_INCREMENET值
mysql> SELECT LAST_INSERT_ID();
+------------------+
| LAST_INSERT_ID() |
+------------------+
|                0 |
+------------------+
1 row in set (0.00 sec)

mysql>
```

5.2.5 加密函数

加密函数是MySQL用来对数据进行加密的函数，以保护数据的安全。

（1）PASSWORD(str)加密函数：可以对字符串str进行加密，采用MySQL的SHA1加密方式，生成的是41位字符串，其中*不加入实际的密码运算。加密函数常用于对用户的密码进行加密。

```
mysql> SELECT PASSWORD('123456');
+-------------------------------------------+
| PASSWORD('123456')                        |
+-------------------------------------------+
| *6BB4837EB74329105EE4568DDA7DC67ED2CA2AD9 |
+-------------------------------------------+
```

（2）MD5(str)加密函数：可以对字符串str进行散列加密，计算字符串str的MD5校验和，常用于一些不需要解密的数据。

```
mysql> SELECT MD5('123456');
+----------------------------------+
| MD5('123456')                    |
+----------------------------------+
| e10adc3949ba59abbe56e057f20f883e |
+----------------------------------+
```

（3）ENCODE(str,pswd_str)与DECODE(crypt_str,pswd_str)加解密函数：是一对加密解密函数，加密结果是二进制数，使用BLOB类型的字段进行存储，使用ENCODE(str,pswd_str)加密函数进行加密，加密后生成解密字符串秘钥，然后使用DECODE(crypt_str,pswd_str)和秘钥进行解密。

```
mysql> SELECT ENCODE('xiaogang','key');
+--------------------+
| ENCODE('xiaogang','key') |
+--------------------+
| 4                  |
+--------------------+
1 row in set, 1 warning (0.00 sec)

mysql> SELECT DECODE(ENCODE('xiaogang','key'),'key');
+-----------------------------+
| DECODE(ENCODE('xiaogang','key'),'key') |
+-----------------------------+
| xiaogang                    |
+-----------------------------+
```

（4）AES_ENCRYPT(str,key)与AES_DECRYPT(str,key)加解密函数：是一对加密解密函数，AES_ENCRYPT(str,key)用秘钥key对字符串利用高级加密标准算法进行加密，加密结果是一个二进制字符串，以BLOB类型的字段进行存储；AES_DECRYPT(str,key)用秘钥key对字符串str利用高级加密标准算法解密。

```
mysql> SELECT AES_ENCRYPT('xiaogang','key');
+------------------------+
| AES_ENCRYPT('xiaogang','key') |
+------------------------+
| y©ë³Æþõ³ñ0             |
+------------------------+
1 row in set (0.00 sec)

mysql> SELECT AES_DECRYPT(AES_ENCRYPT('xiaogang','key'),'key');
+-------------------------------+
| AES_DECRYPT(AES_ENCRYPT('xiaogang','key'),'key') |
+-------------------------------+
| xiaogang                      |
+-------------------------------+
1 row in set (0.00 sec)
```

（5）SHA(str)加解密函数：计算字符串str的安全散列算法(SHA)校验和。

```
mysql> SELECT SHA('123456');
+------------------------+
| SHA('123456')          |
+------------------------+
| 7c4a8d09ca3762af61e59520943dc26494f8941b |
+------------------------+
1 row in set (0.00 sec)
```

5.2.6 格式化函数

MySQL数据库提供了一些格式化函数，它们可以用来格式化日期、时间、浮点数、IP地址。

（1）格式化日期函数DATE_FORMAT(date,fmt)：根据fmt日期格式对date日期进行格式化转换。格式化字符串如表5.5所示。

表5.5 格式化字符串

格式化字符串	含义	示例	示例结果
%M	月份(January…December)	SELECT DATE_FORMAT(NOW(),'%M');	November
%W	星期(Sunday…Saturday)	SELECT DATE_FORMAT(NOW(),'%W');	Thursday
%D	有英语前缀的月份的日期(1st, 2nd, 3rd等)	SELECT DATE_FORMAT(NOW(),'%D');	30th
%Y	年份 数字4位	SELECT DATE_FORMAT(NOW(),'%Y');	2017
%y	年份 数字2位	SELECT DATE_FORMAT(NOW(),'%y');	17
%a	缩写的星期名字(Sun…Sat)	SELECT DATE_FORMAT(NOW(),'%a');	Thu
%d	月份中的天数 数字(00…31)	SELECT DATE_FORMAT(NOW(),'%d');	09
%e	月份中的天数 数字(0…31)	SELECT DATE_FORMAT(NOW(),'%e');	9
%m	月份 数字(01…12)	SELECT DATE_FORMAT(NOW(),'%m');	09
%c	月份 数字(1…12)	SELECT DATE_FORMAT(NOW(),'%c');	9
%b	缩写的月份名字(Jan…Dec)	SELECT DATE_FORMAT(NOW(),'%b');	Sep
%j	一年中的天数(001…366)	SELECT DATE_FORMAT(NOW(),'%j');	244
%H	小时(00…23)，24小时制，2位	SELECT DATE_FORMAT(NOW(),'%H');	08
%k	小时(0…23)，24小时制	SELECT DATE_FORMAT(NOW(),'%k');	8
%h	小时(01…12)，12小时制，2位	SELECT DATE_FORMAT(NOW(),'%h');	08
%I	小时(01…12)，12小时制，2位	SELECT DATE_FORMAT(NOW(),'%I');	08
%l	小时(1…12)，12小时制	SELECT DATE_FORMAT(NOW(),'%l');	8
%i	分钟，数字(00…59)	SELECT DATE_FORMAT(NOW(),'%i');	11
%r	时间,12小时(hh:mm:ss [AP]M)	SELECT DATE_FORMAT(NOW(),'%r');	08：11：31 AM
%T	时间,24小时(hh:mm:ss)	SELECT DATE_FORMAT(NOW(),'%T');	08:11:50
%S	秒(00…59)	SELECT DATE_FORMAT(NOW(),'%S');	06
%s	秒(00…59)	SELECT DATE_FORMAT(NOW(),'%s');	03
%p	AM或PM	SELECT DATE_FORMAT(NOW(),'%p');	AM
%w	一个星期中的天数 (0=Sunday…6=Saturday)	SELECT DATE_FORMAT(NOW(),'%w');	5
%U	星期(0…52)，这里星期天是星期的第一天	SELECT DATE_FORMAT(NOW(),'%U');	35
%u	星期(0…52)，这里星期一是星期的第一天	SELECT DATE_FORMAT(NOW(),'%u');	35
%%	一个文字"%"	SELECT DATE_FORMAT(NOW(),'%%');	%

```
mysql> SELECT DATE_FORMAT(NOW(),'%W,%D %M %Y %r');
+-------------------------------------+
| DATE_FORMAT(NOW(),'%W,%D %M %Y %r') |
+-------------------------------------+
| Thursday,30th November 2017 08：44：29 PM |
+-------------------------------------+
```

```
1 row in set (0.13 sec)

mysql> SELECT DATE_FORMAT(NOW(),'%Y-%m-%d');
+-------------------------------+
| DATE_FORMAT(NOW(),'%Y-%m-%d') |
+-------------------------------+
| 2017-11-30                    |
+-------------------------------+
1 row in set (0.00 sec)

mysql> SELECT DATE_FORMAT(19990330,'%Y-%m-%d');
+----------------------------------+
| DATE_FORMAT(19990330,'%Y-%m-%d') |
+----------------------------------+
| 1999-03-30                       |
+----------------------------------+
1 row in set (0.11 sec)

mysql> SELECT DATE_FORMAT(NOW(),'%h:%i %p');
+-------------------------------+
| DATE_FORMAT(NOW(),'%h:%i %p') |
+-------------------------------+
| 08:44 PM                      |
+-------------------------------+
```

（2）格式化时间函数TIME_FORMAT(time,fmt)：根据fmt时间格式对time时间进行格式化转换。

```
mysql> SELECT TIME_FORMAT(NOW(),'%h %i');
+----------------------------+
| TIME_FORMAT(NOW(),'%h %i') |
+----------------------------+
| 09 33                      |
+----------------------------+
1 row in set (0.00 sec)

mysql> SELECT TIME_FORMAT('21:34:45','%h %i');
+---------------------------------+
| TIME_FORMAT('21:34:45','%h %i') |
+---------------------------------+
| 09 34                           |
+---------------------------------+
1 row in set (0.00 sec)

mysql> SELECT TIME_FORMAT('21:34:45','%h-%i-%s');
+------------------------------------+
| TIME_FORMAT('21:34:45','%h-%i-%s') |
+------------------------------------+
| 09-34-45                           |
+------------------------------------+
1 row in set (0.00 sec)
```

（3）格式化IP函数INET_ATON(ip)和INET_NTOA(num)：INET_ATON(ip)将ip转换为数字，INET_NTOA(num)将数字转换为ip。

```
mysql> SELECT INET_ATON('192.168.1.100');
```

```
| INET_ATON('192.168.1.100') |
+----------------------------+
|                 3232235876 |
+----------------------------+
1 row in set (0.07 sec)

mysql> SELECT INET_NTOA(3232235876);
+-----------------------+
| INET_NTOA(3232235876) |
+-----------------------+
| 192.168.1.100         |
+-----------------------+
1 row in set (0.09 sec)
```

（4）格式化浮点数函数FORMAT(x,y)：把x格式化为以逗号隔开的数字序列，y是结果的小数位数。

```
mysql> SELECT FORMAT(2367.6537,2);
+---------------------+
| FORMAT(2367.6537,2) |
+---------------------+
| 2,367.65            |
+---------------------+
1 row in set (0.13 sec)

mysql> SELECT FORMAT('36828.67628',3);
+-------------------------+
| FORMAT('36828.67628',3) |
+-------------------------+
| 36,828.676              |
+-------------------------+
1 row in set (0.06 sec)
```

5.3 自定义函数

MySQL数据库提供了一些内置的函数，如数学函数、字符串函数等。但是往往在开发过程中，有些业务场景仅靠现有的内置函数并不能满足，这时就可以自定义函数，来满足自己的业务需求。MySQL数据库设计的扩展性允许通过自定义函数的功能来解决这个问题。

5.3.1 函数的基本语法

1. 创建自定义函数

CREATE FUCTION function_name(参数列表)
RETURNS返回值类型
函数体内容

```
mysql> DELIMITER $$
mysql> CREATE FUNCTION fun(a1 varchar(255))
    -> RETURNS VARCHAR(255)
    -> BEGIN
    -> DECLARE x VARCHAR(255) DEFAULT '';
    -> SET x = CONCAT(a1,'word');
    -> RETURN x;
    -> END $$
```

（1）DELIMITER：是用来修改SQL语句的结束符，可以将结束符分号(;)修改为其他特殊字符，因为在函数体内容里可以有多条SQL语句，是用分号(;)隔开的，代表该SQL语句结束，修改后的结束符在函数的末尾添加，代表定义函数完毕，所以要修改结束符。

（2）函数名：使用FUNCTION来标识函数，函数名应该是合法的标识符，不能与已有的关键字或函数冲突，一个函数是属于某数据库的，可以使用db_name.funciton_name的形式执行当前函数所属数据库。

（3）参数列表：可以有0个或者多个参数，每个参数是由参数名和参数类型组成的。

（4）RETURNS：指明函数要返回的类型。

（5）函数体：由多条可用的SQL语句、流程控制语句、变量声明等构成，函数体中一定要含有return返回语句。

（6）BEGIN…END：用来标识函数体的开始和结束，当有多条SQL语句时，可以使用BEGIN…END来标识。

（7）DECLARE：声明变量，包括变量名和类型。

（8）SET：设置变量的值。

（9）RETURN：返回函数体内的返回值。

2. 查看自定义函数

```
SHOW FUNCTION STATUS like 'function_name';
SHOW CREATE FUNCTION function_name;

SHOW FUNCTION STATUS like 'fun';
SHOW CREATE FUNCTION fun;
```

SHOW FUNCTION STATUS用来查看自定义函数的状态，包括所属数据库、类型、函数名称、修改时间等状态的信息；SHOW CREATE FUNCTION用来查看函数的信息，包括函数内容等。

3. 调用自定义函数

```
SELECT function_name(parameter_value,…)

mysql> SELECT FUN();
```

在调用自定义函数的时候，如果有参数，在函数名称的括号里添加参数值。

4. 修改自定义函数

```
ALTER FUNCTION function_name [characteristic …]
characteristic:
{ CONTAINS SQL | NO SQL | READS SQL DATA | MODIFIES SQL DATA }
| SQL SECURITY { DEFINER | INVOKER }
| COMMENT 'string'

mysql> ALTER FUNCTION fun
    -> READS SQL DATA
    -> COMMENT '字符串连接';
```

（1）只能修改自定义函数的一些特性，不能修改函数体，如果要修改函数体，就需要先删除函数来重新创建。

（2）function_name：函数的名称。

（3）characteristic参数：指定存储函数的特性。

（4）CONTAINS SQL：表示子程序包含SQL语句，但不包含读或写数据的语句。

（5）NO SQL：表示子程序中不包含SQL语句。

（6）READS SQL DATA：表示子程序中包含读数据的语句。

（7）MODIFIES SQL DATA：表示子程序中包含写数据的语句。

（8）SQL SECURITY { DEFINER | INVOKER }：指明谁有权限来执行。
（9）DEFINER：表示只有定义者自己才能够执行。
（10）INVOKER：表示调用者可以执行。
（11）COMMENT 'string'：表示注释信息。

5. 删除自定义函数

```
DROP FUNCTION function_name
DROP FUNCTION IF EXISTS function_name
```

mysql> DROP FUNCTION fun;

在删除自定义函数的时候，在FUNCTION后面是函数的名称，不要带括号，也可以使用IF EXISTS判断函数是否存在，然后再进行删除。

5.3.2 创建不带参数的自定义函数

下面完成创建一个日期时间的自定义函数，让它返回2017年12月3日10时48分59秒这样格式的日期时间函数，函数名称为getFormatDatetime。

实战演练——创建不带参数的自定义函数

```
Microsoft Windows [版本 6.1.7601]
版权所有 (c) 2009 Microsoft Corporation。保留所有权利。

C:\Users\Administrator>mysql -uroot -p123456
mysql: [Warning] Using a password on the command line interface can be insecure.

Welcome to the MySQL monitor.  Commands end with ; or \g.
Your MySQL connection id is 4
Server version: 5.7.20 MySQL Community Server (GPL)

Copyright (c) 2000, 2017, Oracle and/or its affiliates. All rights reserved.

Oracle is a registered trademark of Oracle Corporation and/or its
affiliates. Other names may be trademarks of their respective
owners.

Type 'help;' or '\h' for help. Type '\c' to clear the current input statement.

mysql> use shop;
Database changed

#修改默认的结束符";"为"$$",以后的SQL语句都要以"$$"作为结尾
mysql> DELIMITER $$

#创建日期函数getFormatDatetime
mysql> CREATE FUNCTION getFormatDatetime()
    -> RETURNS VARCHAR(255)
    -> RETURN DATE_FORMAT(NOW(),'%Y年%m月%d日%h时%i分%s秒');
    -> $$
Query OK, 0 rows affected (0.00 sec)

#修改结束符为分号(;)
```

```
mysql> DELIMITER ;

#调用日期时间函数
mysql> SELECT getFormatDatetime();
+-------------------+
| getFormatDatetime() |
+-------------------+
| 2017年12月03日10时56分28秒 |
+-------------------+
1 row in set (0.02 sec)

mysql>
```

5.3.3　创建带参数的自定义函数

下面完成创建一个自定义的加法函数，传递两个参数x、y，加法函数完成这两个参数的和，函数名称为addFun。

实战演练——创建带参数的自定义函数

```
Microsoft Windows [版本 6.1.7601]
版权所有 (c) 2009 Microsoft Corporation。保留所有权利。

C:\Users\Administrator>mysql –uroot –p123456
mysql: [Warning] Using a password on the command line interface can be insecure.

Welcome to the MySQL monitor.  Commands end with ; or \g.
Your MySQL connection id is 6
Server version: 5.7.20 MySQL Community Server (GPL)

Copyright (c) 2000, 2017, Oracle and/or its affiliates. All rights reserved.

Oracle is a registered trademark of Oracle Corporation and/or its
affiliates. Other names may be trademarks of their respective
owners.

Type 'help;' or '\h' for help. Type '\c' to clear the current input statement.

mysql> USE shop;
Database changed

#修改结束符
mysql> DELIMITER $$

#创建自定义加法函数addFun
mysql> CREATE FUNCTION addFun(x int UNSIGNED,y int UNSIGNED)
    -> RETURNS int
    -> BEGIN
    -> DECLARE sum int UNSIGNED DEFAULT 0;
    -> SET sum = x + y ;
    -> RETURN sum;
    -> END $$
```

```
Query OK, 0 rows affected (0.00 sec)

#修改结束符
mysql> DELIMITER ;

#调用自定义加法函数
mysql> SELECT addFun(10,26);
+--------------+
| addFun(10,26) |
+--------------+
|           36 |
+--------------+
1 row in set (0.00 sec)

#修改自定义函数
mysql> ALTER FUNCTION addFun
    -> READS SQL DATA
    -> COMMENT '加法自定义函数';
Query OK, 0 rows affected (0.00 sec)

#删除自定义函数
mysql> DROP FUNCTION addFun;
Query OK, 0 rows affected (0.00 sec)

#先判断函数是否存在, 再删除自定义函数
mysql> DROP FUNCTION IF EXISTS addFun;
Query OK, 0 rows affected, 1 warning (0.00 sec)

#如果函数不存在, 没有判断直接删除, 就会报错
mysql> DROP FUNCTION addFun;
ERROR 1305 (42000): FUNCTION shop.addFun does not exist
mysql>
```

5.4 存储过程

存储过程常用于一些复杂的SQL语句的编写, 在开发过程中, 如果不使用存储过程, SQL语句的执行要经过几个步骤: 先校验SQL语句是否正确、语法是否正确, 然后进行SQL语句的编译, 最后再执行SQL语句; 如果复杂语句经过这几个步骤, 运行速度会很慢, 而存储过程就是用来解决这个问题的。存储过程是完成特定功能的SQL语句集, 经编译后存储在数据库中, 在使用的时候直接调用存储过程来执行就可以, 省略了语法校验、编译的过程, 大大提高了SQL语句的执行效率。

精讲视频

存储过程

5.4.1 存储过程的基本语法

1. 创建存储过程

```
CREATE PROCEDURE 过程名([[IN| OUT|INOUT] 参数名 数据类型[,[IN| OUT|INOUT] 参数名 数据类型…]])
[特性 …]
BGIN
过程体
END
```

```
mysql> DELIMITER $$
mysql> CREATE PROCEDURE getUserCountBySex(IN p_sex varchar(255),OUT p_count int)
    -> BEGIN
    -> SELECT COUNT(*) INTO p_count FROM user WHERE sex = p_sex;
    -> END $$
mysql> DELIMITER ;

mysql> SET @p_sex='男';
mysql> SET @p_count=0;
mysql> CALL getUserCountBySex(@p_sex,@p_count);

mysql> SELECT @p_sex,@p_count;
+--------+----------+
| @p_sex | @p_count |
+--------+----------+
| 男     |    4     |
+--------+----------+
```

（1）DELIMITER：是用来修改SQL语句的结束符，可以将结束符分号(;)修改为其他特殊字符，因为在函数体内容里可以有多条SQL语句，是用分号(;)隔开的，代表该SQL语句结束，修改后的结束符在函数的末尾添加，代表定义函数完毕，所以要修改结束符。

（2）过程名：使用PROCEDURE来标识存储过程，存储过程的名称应该是合法的标识符，不能与已有的关键字或存储过程名称冲突。一个存储过程是属于某数据库的，可以使用db_name.procedure_name的形式执行当前存储过程所属的数据库。

（3）参数类型：IN参数的值必须在调用存储过程时指定，在存储过程中修改该参数的值不能被返回；OUT参数的值可在存储过程内部被改变，并可返回；INOUT参数在调用时指定，并且可被改变和返回。

（4）BEGIN…END：用来标识过程体的开始和结束，当有多条SQL语句时，可以使用BEGIN…END来标识。

（5）DECLARE：用来声明变量，包括变量名和类型。

（6）用户变量：一般以@开头，如输入参数变量@p_sex、输出参数变量@p_count，注意滥用用户变量会导致程序难以理解及管理。

（7）SET：用来设置变量的值。

（8）CALL：用来调用存储过程。

（9）存储特性：characteristic参数指定存储函数的特性。

2. 查看存储过程

```
SHOW FUNCTION STATUS like 'procedure_name';
SHOW CREATE FUNCTION procedure_name;

SHOW PROCEDURE STATUS like 'getUserCountBySex';
SHOW CREATE PROCEDURE getUserCountBySex;
```

SHOW PROCEDURE STATUS用来查看自定义存储过程的状态，包括所属数据库、类型、函数名称、修改时间等状态的信息；SHOW CREATE PROCEDURE用来查看存储过程的信息，包括存储过程内容等。

3. 调用存储过程

```
CALL procedure_name(parameter_value,...)

mysql> DELIMITER $$
mysql> CREATE PROCEDURE getUserCountBySex(IN p_sex varchar(255),OUT p_count int)
```

```
        -> BEGIN
        -> DECLARE name VARCHAR(5) DEFAULT '局部变量';
          -> SELECT COUNT(*) INTO p_count FROM user WHERE sex = p_sex;
          -> END $$
mysql> DELIMITER ;

mysql> SET @p_sex='男';
mysql> SET @p_count=0;
mysql> CALL getUserCountBySex(@p_sex,@p_count);
```

在调用自定义存储过程的时候，如果有参数，则在存储过程名称的括号里添加参数值。存储过程变量分为局部变量和用户变量。局部变量在过程体里的BEGIN…END内使用，使用DECLARE关键字定义局部变量，局部变量只在BEGIN…END内有效；而用户变量属于用户客户端变量，与当前用户客户端有关，不同MySQL客户端的用户变量不同，通常使用@符号作为变量的前缀。

4．修改存储过程

```
ALTER  PROCEDURE procedure_name [characteristic ...]
characteristic:
{ CONTAINS SQL | NO SQL | READS SQL DATA | MODIFIES SQL DATA }
| SQL SECURITY { DEFINER | INVOKER }
| COMMENT 'string'

mysql> ALTER PROCEDURE getUserCountBySex
    -> READS SQL DATA
    -> COMMENT '字符串连接';
```

（1）只能修改存储过程的一些特性，不能修改过程体，如果要修改过程体，就需要先删除存储过程来重新创建。

（2）procedure_name：存储过程的名称。

（3）characteristic参数：指定存储函数的特性。

（4）CONTAINS SQL：表示子程序包含SQL语句，但不包含读或写数据的语句。

（5）NO SQL：表示子程序中不包含SQL语句。

（6）READS SQL DATA：表示子程序中包含读数据的语句。

（7）MODIFIES SQL DATA：表示子程序中包含写数据的语句。

（8）SQL SECURITY { DEFINER | INVOKER }：指明谁有权限来执行。

（9）DEFINER：表示只有定义者自己才能够执行。

（10）INVOKER：表示调用者可以执行。

（11）COMMENT 'string'：表示注释信息。

5．删除自定义函数

```
DROP PROCEDURE procedure_name
DROP PROCEDURE IF EXISTS procedure_name

mysql> DROP PROCEDURE getUserCountBySex;
```

在删除存储过程的时候，在PROCEDURE后面是存储过程的名称，不要带括号，也可以使用IF EXISTS来判断存储过程是否存在，然后再进行删除。

5.4.2 创建不带参数的存储过程

可以创建不带参数的存储过程，在调用它的时候可以省略存储过程名称的括号。下面创建一个不带参数的存储过程，来查询用户user的记录。

实战演练——创建不带参数的存储过程

Microsoft Windows [版本 6.1.7601]
版权所有 (c) 2009 Microsoft Corporation。保留所有权利。

C:\Users\Administrator>mysql –uroot –p123456
mysql: [Warning] Using a password on the command line interface can be insecure.

Welcome to the MySQL monitor. Commands end with ; or \g.
Your MySQL connection id is 3
Server version: 5.7.20 MySQL Community Server (GPL)

Copyright (c) 2000, 2017, Oracle and/or its affiliates. All rights reserved.

Oracle is a registered trademark of Oracle Corporation and/or its
affiliates. Other names may be trademarks of their respective
owners.

Type 'help;' or '\h' for help. Type '\c' to clear the current input statement.

mysql> USE shop;
Database changed

mysql> SELETE * FROM user;
```
+----+------+-----+-----+----------+-------+-----------+---------+
| id | name | sex | age | password | phone | loginName | remark  |
+----+------+-----+-----+----------+-------+-----------+---------+
| 3  | david| 女  | 28  | 111111   | NULL  | david     | baann   |
| 4  | 小红 | 女  | 27  | 123456   | NULL  | xiaohong  | black   |
| 5  | 小明 | 男  | 10  | 123456   | NULL  | xiaoming  | berry   |
| 6  | 小刚 | 男  | 12  | 123456   | NULL  | xiaogang  | banner  |
| 7  | 小王 | 男  | 14  | 111111   | NULL  | xiaowang  | banana  |
| 8  | 小绿 | 女  | 34  | 222222   | NULL  | xiaolv    | car     |
| 9  | 晓峰 | 男  | 15  | 333333   | NULL  | xiaofeng  | carray  |
| 10 | 小影 | 女  | 26  | 444444   | NULL  | xiaoying  | baaaa   |
| 11 | 大梅 | 女  | 27  | 555555   | NULL  | damei     | accc    |
+----+------+-----+-----+----------+-------+-----------+---------+
```
9 rows in set (0.00 sec)

#修改结束符$$
mysql> DELIMITER $$

#查询用户表user获取用户信息
mysql> CREATE PROCEDURE getUserInfo()
    -> BEGIN
    -> SELECT * FROM user;
    -> END
    -> $$
Query OK, 0 rows affected (0.06 sec)

#修改结束符

```
mysql> DELIMITER ;
```

#调用存储过程
```
mysql> CALL getUserInfo();
+----+--------+-----+-----+----------+-------+-----------+--------+
| id | name   | sex | age | password | phone | loginName | remark |
+----+--------+-----+-----+----------+-------+-----------+--------+
|  3 | david  | 女  |  28 | 111111   | NULL  | david     | baann  |
|  4 | 小红   | 女  |  27 | 123456   | NULL  | xiaohong  | black  |
|  5 | 小明   | 男  |  10 | 123456   | NULL  | xiaoming  | berry  |
|  6 | 小刚   | 男  |  12 | 123456   | NULL  | xiaogang  | banner |
|  7 | 小王   | 男  |  14 | 111111   | NULL  | xiaowang  | banana |
|  8 | 小绿   | 女  |  34 | 222222   | NULL  | xiaolv    | car    |
|  9 | 晓峰   | 男  |  15 | 333333   | NULL  | xiaofeng  | carray |
| 10 | 小影   | 女  |  26 | 444444   | NULL  | xiaoying  | baaaa  |
| 11 | 大梅   | 女  |  27 | 555555   | NULL  | damei     | accc   |
+----+--------+-----+-----+----------+-------+-----------+--------+
9 rows in set (1.01 sec)

Query OK, 0 rows affected (1.10 sec)
```

#不带参数的存储过程，在调用的时候可以省略括号
```
mysql> CALL getUserInfo;
+----+--------+-----+-----+----------+-------+-----------+--------+
| id | name   | sex | age | password | phone | loginName | remark |
+----+--------+-----+-----+----------+-------+-----------+--------+
|  3 | david  | 女  |  28 | 111111   | NULL  | david     | baann  |
|  4 | 小红   | 女  |  27 | 123456   | NULL  | xiaohong  | black  |
|  5 | 小明   | 男  |  10 | 123456   | NULL  | xiaoming  | berry  |
|  6 | 小刚   | 男  |  12 | 123456   | NULL  | xiaogang  | banner |
|  7 | 小王   | 男  |  14 | 111111   | NULL  | xiaowang  | banana |
|  8 | 小绿   | 女  |  34 | 222222   | NULL  | xiaolv    | car    |
|  9 | 晓峰   | 男  |  15 | 333333   | NULL  | xiaofeng  | carray |
| 10 | 小影   | 女  |  26 | 444444   | NULL  | xiaoying  | baaaa  |
| 11 | 大梅   | 女  |  27 | 555555   | NULL  | damei     | accc   |
+----+--------+-----+-----+----------+-------+-----------+--------+
9 rows in set (0.00 sec)

Query OK, 0 rows affected (0.11 sec)

mysql>
```

5.4.3 创建带有IN类型参数的存储过程

IN类型参数作为输入参数，在存储过程调用时指定。IN类型参数在存储过程中修改后该参数的值不能被返回。下面创建一个带有IN类型参数的存储过程，传入用户的性别，来查询满足条件的用户user的记录。

实战演练——创建带有IN类型参数的存储过程

```
Microsoft Windows [版本 6.1.7601]
版权所有 (c) 2009 Microsoft Corporation。保留所有权利。

C:\Users\Administrator>mysql -uroot -p123456
```

```
mysql: [Warning] Using a password on the command line interface can be insecure.

Welcome to the MySQL monitor.  Commands end with ; or \g.
Your MySQL connection id is 4
Server version: 5.7.20 MySQL Community Server (GPL)

Copyright (c) 2000, 2017, Oracle and/or its affiliates. All rights reserved.

Oracle is a registered trademark of Oracle Corporation and/or its
affiliates. Other names may be trademarks of their respective
owners.

Type 'help;' or '\h' for help. Type '\c' to clear the current input statement.
```

mysql> USE shop;
Database changed

mysql> SELETE * FROM user;
```
+----+-------+-----+-----+----------+-------+-----------+--------+
| id | name  | sex | age | password | phone | loginName | remark |
+----+-------+-----+-----+----------+-------+-----------+--------+
|  3 | david | 女  |  28 | 111111   | NULL  | david     | baann  |
|  4 | 小红  | 女  |  27 | 123456   | NULL  | xiaohong  | black  |
|  5 | 小明  | 男  |  10 | 123456   | NULL  | xiaoming  | berry  |
|  6 | 小刚  | 男  |  12 | 123456   | NULL  | xiaogang  | banner |
|  7 | 小王  | 男  |  14 | 111111   | NULL  | xiaowang  | banana |
|  8 | 小绿  | 女  |  34 | 222222   | NULL  | xiaolv    | car    |
|  9 | 晓峰  | 男  |  15 | 333333   | NULL  | xiaofeng  | carray |
| 10 | 小影  | 女  |  26 | 444444   | NULL  | xiaoying  | baaaa  |
| 11 | 大梅  | 女  |  27 | 555555   | NULL  | damei     | accc   |
+----+-------+-----+-----+----------+-------+-----------+--------+
9 rows in set (0.00 sec)
```

#修改结束符$$
mysql> DELIMITER $$

#创建IN类型参数的存储过程，根据性别查询用户记录
mysql> CREATE PROCEDURE getUserBySex(IN p_sex varchar(255))
    -> BEGIN
    -> SELECT p_sex;
    -> SELECT * FROM user WHERE sex = p_sex;
    -> SET p_sex = '未知';
    -> SELECT p_sex;
    -> END
    -> $$
Query OK, 0 rows affected (0.00 sec)

#修改结束符
mysql> DELIMITER ;

#调用存储过程，输入参数可以直接在括号里赋值

```
mysql> CALL getUserBySex('女');
+-------+
| p_sex |
+-------+
| 女    |
+-------+
1 row in set (0.00 sec)

+----+------+-----+-----+----------+-------+-----------+--------+
| id | name | sex | age | password | phone | loginName | remark |
+----+------+-----+-----+----------+-------+-----------+--------+
|  3 | david| 女  | 28  | 111111   | NULL  | david     | baann  |
|  4 | 小红 | 女  | 27  | 123456   | NULL  | xiaohong  | black  |
|  8 | 小绿 | 女  | 34  | 222222   | NULL  | xiaolv    | car    |
| 10 | 小影 | 女  | 26  | 444444   | NULL  | xiaoying  | baaaa  |
| 11 | 大梅 | 女  | 27  | 555555   | NULL  | damei     | accc   |
+----+------+-----+-----+----------+-------+-----------+--------+
5 rows in set (0.05 sec)

+-------+
| p_sex |
+-------+
| 未知  |
+-------+
1 row in set (0.11 sec)

Query OK, 0 rows affected (0.12 sec)
```

#IN类型参数，也可以通过定义用户变量@p_sex来传递参数
```
mysql> SET @p_sex='男';
Query OK, 0 rows affected (0.00 sec)
```

#传入用户变量，调用存储过程
```
mysql> CALL getUserBySex(@p_sex);
+-------+
| p_sex |
+-------+
| 男    |
+-------+
1 row in set (0.00 sec)

+----+------+-----+-----+----------+-------+-----------+--------+
| id | name | sex | age | password | phone | loginName | remark |
+----+------+-----+-----+----------+-------+-----------+--------+
|  5 | 小明 | 男  | 10  | 123456   | NULL  | xiaoming  | berry  |
|  6 | 小刚 | 男  | 12  | 123456   | NULL  | xiaogang  | banner |
|  7 | 小王 | 男  | 14  | 111111   | NULL  | xiaowang  | banana |
|  9 | 晓峰 | 男  | 15  | 333333   | NULL  | xiaofeng  | carray |
+----+------+-----+-----+----------+-------+-----------+--------+
4 rows in set (0.00 sec)
```

```
| p_sex |
+-----+
| 未知  |
+-----+
1 row in set (0.06 sec)

Query OK, 0 rows affected (0.07 sec)

#在存储过程中修改IN类型输入参数，但是并不能返回，所以@p_sex还是修改前的值
mysql> SELECT @p_sex;
+-----+
| @p_sex |
+-----+
| 男    |
+-----+
1 row in set (0.00 sec)

mysql>
```

5.4.4　创建带有IN和OUT类型参数的存储过程

存储过程中使用IN作为输入参数的类型标识，OUT作为输出参数的类型标识，可以同时传入输入、输出参数。IN类型输入参数的值在存储过程中修改后不能被返回，OUT类型输出参数的值在存储过程中修改后可以被返回。下面创建一个带有IN和OUT类型参数的存储过程，传入用户的性别，返回该性别的用户数量。

实战演练——创建带有IN和OUT类型参数的存储过程

```
Microsoft Windows [版本 6.1.7601]
版权所有 (c) 2009 Microsoft Corporation。保留所有权利。

C:\Users\Administrator>mysql –uroot –p123456
mysql: [Warning] Using a password on the command line interface can be insecure.

Welcome to the MySQL monitor.  Commands end with ; or \g.
Your MySQL connection id is 7
Server version: 5.7.20 MySQL Community Server (GPL)

Copyright (c) 2000, 2017, Oracle and/or its affiliates. All rights reserved.

Oracle is a registered trademark of Oracle Corporation and/or its
affiliates. Other names may be trademarks of their respective
owners.

Type 'help;' or '\h' for help. Type '\c' to clear the current input statement.

mysql> USE shop;
Database changed

mysql> SELETE * FROM user;
+--+----+--+---+--------+----+-----+------+
| id | name | sex | age | password | phone | loginName | remark |
```

```
+--+----+----+----+------+------+---------+--------+
| 3|david| 女 | 28 |111111| NULL | david   | baann  |
| 4|小红 | 女 | 27 |123456| NULL | xiaohong| black  |
| 5|小明 | 男 | 10 |123456| NULL | xiaoming| berry  |
| 6|小刚 | 男 | 12 |123456| NULL | xiaogang| banner |
| 7|小王 | 男 | 14 |111111| NULL | xiaowang| banana |
| 8|小绿 | 女 | 34 |222222| NULL | xiaolv  | car    |
| 9|晓峰 | 男 | 15 |333333| NULL | xiaofeng| carray |
|10|小影 | 女 | 26 |444444| NULL | xiaoying| baaaa  |
|11|大梅 | 女 | 27 |555555| NULL | damei   | accc   |
+--+----+----+----+------+------+---------+--------+
```
9 rows in set (0.00 sec)

#修改结束符
mysql> DELIMITER $$
mysql> CREATE PROCEDURE getCountBySex(IN p_sex varchar(255),OUT p_count int)
　　-> BEGIN
　　-> SELECT COUNT(*) INTO p_count FROM user WHERE sex=p_sex;
　　-> END
　　-> $$
Query OK, 0 rows affected (0.00 sec)

#修改结束符
mysql> DELIMITER ;

#传入输入参数，定义输出参数@p_count，调用存储过程
mysql> CALL getCountBySex('女',@p_count);
Query OK, 1 row affected (0.07 sec)

#查询输出参数@p_sex
mysql> SELECT @p_count;
```
+------+
|@p_sex|
+------+
|   5  |
+------+
```
1 row in set (0.00 sec)

#定义输入参数@p_sex
mysql> SET @p_sex='男';
Query OK, 0 rows affected (0.00 sec)

#定义输出参数@p_count
mysql> SET @p_count=0;
Query OK, 0 rows affected (0.00 sec)

#调用存储过程
mysql> CALL getCountBySex(@p_sex,@p_count);
Query OK, 1 row affected (0.00 sec)

#查询输出结果
mysql> SELECT @p_count;

```
+----------+
| @p_count |
+----------+
|        4 |
+----------+
1 row in set (0.00 sec)

mysql>
```

5.4.5 创建带有多个OUT类型参数的存储过程

存储过程不仅可以输入一个IN类型的参数、输出一个OUT类型的参数，也可以输入、输出多个参数，存储过程可以创建带有多个OUT类型的输出参数的存储过程，也就是说可以有多个输出参数。下面创建一个带有多个OUT类型参数的存储过程，输入id的值，来查询小于该id的男生数量和女生数量分别是多少。

实战演练——创建带有多个OUT类型参数的存储过程

Microsoft Windows [版本 6.1.7601]
版权所有 (c) 2009 Microsoft Corporation。保留所有权利。

C:\Users\Administrator>mysql –uroot –p123456
mysql: [Warning] Using a password on the command line interface can be insecure.

Welcome to the MySQL monitor. Commands end with ; or \g.
Your MySQL connection id is 8
Server version: 5.7.20 MySQL Community Server (GPL)

Copyright (c) 2000, 2017, Oracle and/or its affiliates. All rights reserved.

Oracle is a registered trademark of Oracle Corporation and/or its
affiliates. Other names may be trademarks of their respective
owners.

Type 'help;' or '\h' for help. Type '\c' to clear the current input statement.

mysql> USE shop;
Database changed

```
mysql> SELECT * FROM user;
+----+------+-----+-----+----------+-------+-----------+--------+
| id | name | sex | age | password | phone | loginName | remark |
+----+------+-----+-----+----------+-------+-----------+--------+
|  3 | david| 女  |  28 | 111111   | NULL  | david     | baann  |
|  4 | 小红 | 女  |  27 | 123456   | NULL  | xiaohong  | black  |
|  5 | 小明 | 男  |  10 | 123456   | NULL  | xiaoming  | berry  |
|  6 | 小刚 | 男  |  12 | 123456   | NULL  | xiaogang  | banner |
|  7 | 小王 | 男  |  14 | 111111   | NULL  | xiaowang  | banana |
|  8 | 小绿 | 女  |  34 | 222222   | NULL  | xiaolv    | car    |
|  9 | 晓峰 | 男  |  15 | 333333   | NULL  | xiaofeng  | carray |
| 10 | 小影 | 女  |  26 | 444444   | NULL  | xiaoying  | baaaa  |
| 11 | 大梅 | 女  |  27 | 555555   | NULL  | damei     | accc   |
+----+------+-----+-----+----------+-------+-----------+--------+
```

9 rows in set (0.00 sec)

```
#修改结束符
mysql> DELIMITER $$

#创建存储过程,输入一个参数,输出两个参数
mysql> CREATE PROCEDURE getSexCountById(IN p_id int,OUT p_m_count int,OUT p_f_count int)
    -> BEGIN
    -> SELECT COUNT(*) INTO p_m_count FROM USER WHERE id < p_id and sex='男';
    -> SELECT COUNT(*) INTO p_f_count FROM USER WHERE id < p_id and sex='女';
    -> END
    -> $$
Query OK, 0 rows affected (0.00 sec)

#修改结束符
mysql> DELIMITER ;

#调用存储过程
mysql> CALL getSexCountById(8,@p_m_count,@p_f_count);
Query OK, 1 row affected (0.00 sec)

#输出性别为男的数量
mysql> SELECT @p_m_count;
+------------+
| @p_m_count |
+------------+
|          3 |
+------------+
1 row in set (0.00 sec)

#输出性别为女的数量
mysql> SELECT @p_f_count;
+------------+
| @p_f_count |
+------------+
|          2 |
+------------+
1 row in set (0.00 sec)

mysql>
```

5.4.6　创建带有INOUT类型参数的存储过程

　　INOUT类型参数调用时指定参数,该类型参数的值可以被改变以及返回。下面创建一个带有INOUT类型参数的存储过程,根据用户的id查询用户user记录。

实战演练——创建带有INOUT类型参数的存储过程

Microsoft Windows [版本 6.1.7601]
版权所有 (c) 2009 Microsoft Corporation。保留所有权利。

C:\Users\Administrator>mysql -uroot -p123456

```
mysql: [Warning] Using a password on the command line interface can be insecure.

Welcome to the MySQL monitor.  Commands end with ; or \g.
Your MySQL connection id is 9
Server version: 5.7.20 MySQL Community Server (GPL)

Copyright (c) 2000, 2017, Oracle and/or its affiliates. All rights reserved.

Oracle is a registered trademark of Oracle Corporation and/or its
affiliates. Other names may be trademarks of their respective
owners.

Type 'help;' or '\h' for help. Type '\c' to clear the current input statement.
```

mysql> USE shop;
Database changed

mysql> SELECT * FROM user;

```
+----+-------+-----+-----+----------+-------+-----------+--------+
| id | name  | sex | age | password | phone | loginName | remark |
+----+-------+-----+-----+----------+-------+-----------+--------+
|  3 | david | 女  |  28 | 111111   | NULL  | david     | baann  |
|  4 | 小红  | 女  |  27 | 123456   | NULL  | xiaohong  | black  |
|  5 | 小明  | 男  |  10 | 123456   | NULL  | xiaoming  | berry  |
|  6 | 小刚  | 男  |  12 | 123456   | NULL  | xiaogang  | banner |
|  7 | 小王  | 男  |  14 | 111111   | NULL  | xiaowang  | banana |
|  8 | 小绿  | 女  |  34 | 222222   | NULL  | xiaolv    | car    |
|  9 | 晓峰  | 男  |  15 | 333333   | NULL  | xiaofeng  | carray |
| 10 | 小影  | 女  |  26 | 444444   | NULL  | xiaoying  | baaaa  |
| 11 | 大梅  | 女  |  27 | 555555   | NULL  | damei     | accc   |
+----+-------+-----+-----+----------+-------+-----------+--------+
9 rows in set (0.00 sec)
```

#修改结束符
mysql> DELIMITER $$

#创建存储过程，传入INOUT类型的用户id，根据用户id查找用户的记录
mysql> CREATE PROCEDURE getUserById(INOUT p_id int)
　　-> BEGIN
　　-> SELECT p_id;
　　-> SELECT * FROM user WHERE id = p_id;
　　-> SET p_id =1000;
　　-> SELECT p_id;
　　-> END
　　-> $$
Query OK, 0 rows affected (0.05 sec)

#修改结束符
mysql> DELIMITER ;

#设置输入变量的值为8
mysql> SET @p_id = 8;

```
Query OK, 0 rows affected (0.00 sec)

#调用存储过程
mysql> CALL getUserById(@p_id);
+------+
| p_id |
+------+
|    8 |
+------+
1 row in set (0.00 sec)

+----+------+-----+-----+----------+-------+-----------+--------+
| id | name | sex | age | password | phone | loginName | remark |
+----+------+-----+-----+----------+-------+-----------+--------+
|  8 | 小绿 | 女  |  34 | 222222   | NULL  | xiaolv    | car    |
+----+------+-----+-----+----------+-------+-----------+--------+
1 row in set (0.01 sec)

+------+
| p_id |
+------+
| 1000 |
+------+
1 row in set (0.04 sec)

Query OK, 0 rows affected (0.04 sec)

#INOUT类型的参数值可以被改变，输入时是8，在存储过程中被重新赋值后变为1000
mysql> SELECT @p_id;
+-------+
| @p_id |
+-------+
|  1000 |
+-------+
1 row in set (0.00 sec)

mysql>
```

5.4.7 创建IF语句的存储过程

在存储过程中可以使用IF语句来进行条件判断，这也是经常会用到的控制语句。下面创建一个存储过程，判断用户的年龄，0~6岁时输出童年，7~17岁时输出少年，18~40岁时输出青年。

实战演练——创建IF语句的存储过程

```
Microsoft Windows [版本 6.1.7601]
版权所有 (c) 2009 Microsoft Corporation。保留所有权利。

C:\Users\Administrator>mysql -uroot -p123456
mysql: [Warning] Using a password on the command line interface can be insecure.

Welcome to the MySQL monitor.  Commands end with ; or \g.
Your MySQL connection id is 10
```

```
Server version: 5.7.20 MySQL Community Server (GPL)

Copyright (c) 2000, 2017, Oracle and/or its affiliates. All rights reserved.

Oracle is a registered trademark of Oracle Corporation and/or its
affiliates. Other names may be trademarks of their respective
owners.

Type 'help;' or '\h' for help. Type '\c' to clear the current input statement.

mysql> USE shop;
Database changed

#修改结束符
mysql> DELIMITER $$

#输入用户的年龄，查询用户年龄所属的类型
mysql> CREATE PROCEDURE getTypeByAge(IN p_age int)
    -> BEGIN
    -> DECLARE type varchar(255) DEFAULT '';
    -> IF (p_age >0 AND p_age <=6) THEN
    -> SET type = '童年';
    -> ELSEIF(p_age>6 AND p_age <=17) THEN
    -> SET type = '少年';
    -> ELSE
    -> SET type ='成年';
    -> END IF;
    -> SELECT type;
    -> END
    -> $$
Query OK, 0 rows affected (0.00 sec)

#修改结束符
mysql> DELIMITER ;

#输入年龄为5，年龄类型为童年
mysql> CALL getTypeByAge(5);
+------+
| type |
+------+
| 童年 |
+------+
1 row in set (0.00 sec)

Query OK, 0 rows affected (0.01 sec)

#输入年龄为14，年龄类型为少年
mysql> CALL getTypeByAge(14);
+------+
| type |
+------+
| 少年 |
```

```
+-----+
1 row in set (0.00 sec)

Query OK, 0 rows affected (0.01 sec)

#输入年龄为25，年龄类型为成年
mysql> CALL getTypeByAge(25);
+-----+
| type |
+-----+
| 成年 |
+-----+
1 row in set (0.00 sec)

Query OK, 0 rows affected (0.01 sec)

mysql>
```

5.4.8 创建CASE语句的存储过程

存储过程可以使用CASE语句来进行条件判断。下面创建一个CASE语句的存储过程，根据传入的类型，输出用户记录的条数，输入0时输出2条记录，输入1时输出5条记录，不满足这两个条件时输出所有记录。

实战演练——创建CASE语句的存储过程

```
Microsoft Windows [版本 6.1.7601]
版权所有 (c) 2009 Microsoft Corporation。保留所有权利。

C:\Users\Administrator>mysql -uroot -p123456
mysql: [Warning] Using a password on the command line interface can be insecure.

Welcome to the MySQL monitor.  Commands end with ; or \g.
Your MySQL connection id is 11
Server version: 5.7.20 MySQL Community Server (GPL)

Copyright (c) 2000, 2017, Oracle and/or its affiliates. All rights reserved.

Oracle is a registered trademark of Oracle Corporation and/or its
affiliates. Other names may be trademarks of their respective
owners.

Type 'help;' or '\h' for help. Type '\c' to clear the current input statement.

mysql> USE shop;
Database changed

mysql> SELECT * FROM user;
+----+------+-----+-----+--------+-------+-----------+--------+
| id | name | sex | age | password | phone | loginName | remark |
+----+------+-----+-----+--------+-------+-----------+--------+
| 3  | david| 女  | 28  | 111111 | NULL  | david     | baann  |
| 4  | 小红 | 女  | 27  | 123456 | NULL  | xiaohong  | black  |
| 5  | 小明 | 男  | 10  | 123456 | NULL  | xiaoming  | berry  |
```

```
|  6 | 小刚 | 男 | 12 | 123456 | NULL | xiaogang | banner |
|  7 | 小王 | 男 | 14 | 111111 | NULL | xiaowang | banana |
|  8 | 小绿 | 女 | 34 | 222222 | NULL | xiaolv   | car    |
|  9 | 晓峰 | 男 | 15 | 333333 | NULL | xiaofeng | carray |
| 10 | 小影 | 女 | 26 | 444444 | NULL | xiaoying | baaaa  |
| 11 | 大梅 | 女 | 27 | 555555 | NULL | damei    | accc   |
+----+------+----+----+--------+------+----------+--------+
9 rows in set (0.00 sec)
```

#修改结束符
mysql> DELIMITER $$

#使用CASE语句判断输入类型的值，输入0时输出2条记录，输入1时输出5条记录，不满足这两个条件时输出所有记录
```
mysql> CREATE PROCEDURE getUserByType(IN p_type int)
    -> BEGIN
    -> CASE p_type
    -> WHEN 0 THEN
    -> SELECT * FROM user LIMIT 0,2;
    -> WHEN 1 THEN
    -> SELECT * FROM user LIMIT 0,5;
    -> ELSE
    -> SELECT * FROM user;
    -> END CASE;
    -> END
    -> $$
Query OK, 0 rows affected (0.00 sec)
```

#修改结束符
mysql> DELIMITER ;

#输入0，输出2条记录
```
mysql> CALL getUserByType(0);
+----+------+----+-----+--------+-------+----------+--------+
| id | name | sex | age | password | phone | loginName | remark |
+----+------+----+-----+--------+-------+----------+--------+
|  3 | david | 女 | 28 | 111111 | NULL | david    | baann  |
|  4 | 小红  | 女 | 27 | 123456 | NULL | xiaohong | black  |
+----+------+----+-----+--------+-------+----------+--------+
2 rows in set (0.00 sec)

Query OK, 0 rows affected (0.06 sec)
```

#输入1，输出5条记录
```
mysql> CALL getUserByType(1);
+----+------+----+-----+--------+-------+----------+--------+
| id | name | sex | age | password | phone | loginName | remark |
+----+------+----+-----+--------+-------+----------+--------+
|  3 | david | 女 | 28 | 111111 | NULL | david    | baann  |
|  4 | 小红  | 女 | 27 | 123456 | NULL | xiaohong | black  |
|  5 | 小明  | 男 | 10 | 123456 | NULL | xiaoming | berry  |
|  6 | 小刚  | 男 | 12 | 123456 | NULL | xiaogang | banner |
```

```
| 7 | 小王  | 男  | 14 | 111111 | NULL | xiaowang | banana |
+---+------+----+----+--------+------+----------+--------+
5 rows in set (0.00 sec)

Query OK, 0 rows affected (0.13 sec)


#不满足条件时，输出所有记录
mysql> CALL getUserByType(2);
+----+-------+-----+-----+--------+-------+----------+--------+
| id | name  | sex | age | password | phone | loginName | remark |
+----+-------+-----+-----+--------+-------+----------+--------+
|  3 | david | 女  | 28  | 111111 | NULL  | david    | baann  |
|  4 | 小红  | 女  | 27  | 123456 | NULL  | xiaohong | black  |
|  5 | 小明  | 男  | 10  | 123456 | NULL  | xiaoming | berry  |
|  6 | 小刚  | 男  | 12  | 123456 | NULL  | xiaogang | banner |
|  7 | 小王  | 男  | 14  | 111111 | NULL  | xiaowang | banana |
|  8 | 小绿  | 女  | 34  | 222222 | NULL  | xiaolv   | car    |
|  9 | 晓峰  | 男  | 15  | 333333 | NULL  | xiaofeng | carray |
| 10 | 小影  | 女  | 26  | 444444 | NULL  | xiaoying | baaaa  |
| 11 | 大梅  | 女  | 27  | 555555 | NULL  | damei    | accc   |
+----+-------+-----+-----+--------+-------+----------+--------+
9 rows in set (0.00 sec)

Query OK, 0 rows affected (0.20 sec)

mysql>
```

5.4.9 创建while循环语句的存储过程

while循环语句也是在存储过程中经常会用到的语句。下面创建一个存储过程，根据输入参数的值，计算累加和。

实战演练——创建while循环语句的存储过程

Microsoft Windows [版本 6.1.7601]
版权所有 (c) 2009 Microsoft Corporation。保留所有权利。

C:\Users\Administrator>mysql -uroot -p123456
mysql: [Warning] Using a password on the command line interface can be insecure.

Welcome to the MySQL monitor. Commands end with ; or \g.
Your MySQL connection id is 12
Server version: 5.7.20 MySQL Community Server (GPL)

Copyright (c) 2000, 2017, Oracle and/or its affiliates. All rights reserved.

Oracle is a registered trademark of Oracle Corporation and/or its
affiliates. Other names may be trademarks of their respective
owners.

Type 'help;' or '\h' for help. Type '\c' to clear the current input statement.

```
mysql> USE shop;
Database changed

#修改结束符
mysql> DELIMITER $$

#使用while循环语句，计算累加和
mysql> CREATE PROCEDURE computeSum(IN n int)
    -> BEGIN
    -> DECLARE i int DEFAULT 0;
    -> DECLARE s int DEFAULT 0;
    -> WHILE i <= n DO
    -> SET s = s + i;
    -> SET i = i + 1;
    -> END WHILE;
    -> SELECT s;
    -> END
    -> $$
Query OK, 0 rows affected (0.00 sec)

#修改结束符
mysql> DELIMITER ;

#计算0-100累加和
mysql> CALL computeSum(100);
+------+
| s    |
+------+
| 5050 |
+------+
1 row in set (0.00 sec)

Query OK, 0 rows affected (0.01 sec)

mysql>
```

5.5 自定义函数和存储过程的区别

5.5.1 自定义函数和存储过程的区别

自定义函数可以作为内置函数的扩展使用，针对性比较强，作为SQL语句的一部分来使用，并且只有一个返回值；存储过程是为了解决复杂的SQL语句或业务逻辑，预编译保存在数据库中，使用的时候直接从数据库中调用，这样可以提高执行效率，它可以有多个返回值，可以独立运行，同时它可以降低网络的数据传输量。

精讲视频

自定义函数和存
储过程的区别

5.5.2 存储过程的使用建议

存储过程中可以将SQL语句预编译保存在数据库中，使用的时候直接调用，这大大提高了执行效率，同时降低了网络数据传输量，这是它的好处。那么，存储过程是不是可以大量使用呢？

在使用存储过程的时候应该注意以下几点。

（1）各个版本的数据库（MySQL、Oracle、SQL Server等）在存储过程中语法有可能不一样，不利于数据库的移植。

（2）版本不好控制，不能进行多人协同开发，调试不方便，同时要掌握对存储过程的开发，才能写出比较好的、效率高的存储过程。

（3）将复杂的核心业务逻辑放在存储过程中，一旦业务发生改变或者业务逻辑频繁变化，存储过程就需要重新编写、调试，花费代价很大，所以应该把核心业务或者经常变化的功能放在程序中来处理，而不应该放在存储过程中。

（4）对效率要求非常高的时候可以使用少量的存储过程。

5.6 小结

本章主要讲述了MySQL函数和存储过程的使用。学习完本章后，要掌握流程控制函数的使用、MySQL内置函数的使用；学会如何自定义函数、创建不带参数的自定义函数、创建带参数的自定义函数；学会存储过程的使用及创建存储过程，理解IN、OUT、INOUT参数类型的含义以及其参数的传递，同时要学会IF、CASE、while语句在存储过程中的使用；最后了解自定义函数和存储过程的区别，理解其使用场景。

第6章
MySQL 高级特性

本章要点

- 视图
- 游标
- 触发器
- 小结

本章讲解MySQL的高级特性,包括创建视图,修改、更新、删除视图数据,使用游标,在游标中使用WHILE、REPEAT、LOOP循环,创建、查看、删除触发器以及INSERT型、UPDATE型、DELETE型触发器的使用。这些高级特性能帮助我们更高效地应用MySQL数据库。

6.1 视图

6.1.1 什么是视图

视图

视图是一条SELECT语句返回的结果集,这个结果集可以从一张表中查询出来,也可以从多张表中查询出来。SELECT语句使用的表可以当成基本表,而结果集则构成一张虚拟表。虚拟表也是表,可以进行增、删、改、查操作,但是有条件限制,它可以存放SELECT语句查询的结果,但是不存放具体数据,基本表里的数据变化会影响视图虚拟表查询的结果。

视图的好处在于方便查询,只查询出想要的字段。假如有三张表,每张表有两个字段是需要作为返回结果的,这时可以创建一个视图,把这三张表的两个字段都放在视图虚拟表中,查询的时候直接查询视图,可以获取到这三张表的两个字段数据,方便SQL语句查询,减少复杂的SQL语句,增强可读性。同时,使用视图也更加安全,基本表查询不能限定特定的列和行,而有不想显示出来的存在敏感信息的列时,视图对其可以灵活地控制。

6.1.2 创建视图

创建视图使用关键字VIEW来标识。
```
CREATE [ALGORITHM]={UNDEFINED|MERGE|TEMPTABLE}]
    VIEW 视图名 [属性清单]
    AS SELECT 语句
    [WITH [CASCADED|LOCAL] CHECK OPTION];
CREATE VIEW user_view AS SELECT id,name FROM user WHERE id=1;
```
(1)可选参数ALGORITHM代表视图选择的算法,UNDEFINED参数值代表将自动选择所要使用的算法,MERGE参数值代表将视图的语句与视图定义合并起来, TEMPTABLE参数值将视图的结果存入临时表,然后使用临时表执行语句。

(2)视图名类似于表名,给一个视图命名,查询该视图时就要用到视图名。

(3)属性清单代表视图中的列名,默认其与SELECT查询结果中的列名相同,也可以重新自定义列名。

(4)可选参数WITH CHECK OPTION代表更新视图时在权限范围内。

(5)可选参数CASCADED代表更新视图时要满足所有相关视图和表的条件。

(6)可选参数LOCAL代表更新视图时要满足该视图本身定义的条件。

实战演练——创建视图

下面创建一个员工表employee,包含字段主键员工id、姓名name、性别sex、年龄age、部门外键deptId;创建一个部门表dept,包含字段主键部门id、部门名称deptName、部门经理manager;创建一个工资表payroll,包含字段主键工资id、员工外键empId、工资salary、发放日期grantDate;基于员工表、部门表、工资表创建一个视图,用来查询员工id、姓名name、部门名称deptName、工资salary、发放日期grantDate。

```
Microsoft Windows [版本 6.1.7601]
版权所有 (c) 2009 Microsoft Corporation。保留所有权利。

C:\Users\Administrator>mysql -uroot -p123456
mysql: [Warning] Using a password on the command line interface can be insecure.

Welcome to the MySQL monitor.  Commands end with ; or \g.
```

```
Your MySQL connection id is 12
Server version: 5.7.20 MySQL Community Server (GPL)

Copyright (c) 2000, 2017, Oracle and/or its affiliates. All rights reserved.

Oracle is a registered trademark of Oracle Corporation and/or its
affiliates. Other names may be trademarks of their respective
owners.

Type 'help;' or '\h' for help. Type '\c' to clear the current input statement.

mysql> USE shop
Database changed

#创建员工表employee
mysql> CREATE TABLE employee(
    -> id int not null AUTO_INCREMENT,
    -> name varchar(255),
    -> sex varchar(10),
    -> age int,
    -> deptId int,
    -> primary key(id)
    -> );
Query OK, 0 rows affected (0.34 sec)

#创建部门表dept
mysql> CREATE TABLE dept(
    -> id int not null AUTO_INCREMENT,
    -> deptName varchar(255),
    -> manager varchar(255),
    -> primary key(id)
    -> );
Query OK, 0 rows affected (0.38 sec)

#创建工资表payroll
mysql> CREATE TABLE payroll(
    -> id int not null AUTO_INCREMENT,
    -> empId int,
    -> salary varchar(255),
    -> grantDate date,
    -> primary key(id)
    -> );
Query OK, 0 rows affected (0.46 sec)

#插入员工
mysql> INSERT INTO employee VALUES(1,'张明','男',30,1);
Query OK, 1 row affected (0.08 sec)

#插入员工
mysql> INSERT INTO employee VALUES(2,'孙浩','男',25,1);
Query OK, 1 row affected (0.00 sec)
```

```
#插入员工
mysql> INSERT INTO employee VALUES(3,'张静','女',28,2);
Query OK, 1 row affected (0.00 sec)

#插入员工
mysql> INSERT INTO employee VALUES(4,'赵颖','女',32,2);
Query OK, 1 row affected (0.00 sec)

#插入员工
mysql> INSERT INTO employee VALUES(5,'刘帅','男',28,2);
Query OK, 1 row affected (0.00 sec)

#插入部门
mysql> INSERT INTO dept VALUES(1,'软件开发部','王洋');
Query OK, 1 row affected (0.00 sec)

#插入部门
mysql> INSERT INTO dept VALUES(2,'人力资源部','吴刚');
Query OK, 1 row affected (0.00 sec)

#插入工资
mysql> INSERT INTO payroll VALUES(1,1,'13500','2017-11-15');
Query OK, 1 row affected (0.00 sec)

#插入工资
mysql> INSERT INTO payroll VALUES(2,1,'16500','2017-12-15');
Query OK, 1 row affected (0.00 sec)

#插入工资
mysql> INSERT INTO payroll VALUES(3,2,'9500','2017-11-15');
Query OK, 1 row affected (0.00 sec)

#插入工资
mysql> INSERT INTO payroll VALUES(4,2,'10500','2017-12-15');
Query OK, 1 row affected (0.00 sec)

#插入工资
mysql> INSERT INTO payroll VALUES(5,3,'11000','2017-11-15');
Query OK, 1 row affected (0.00 sec)

#插入工资
mysql> INSERT INTO payroll VALUES(6,3,'12000','2017-12-15');
Query OK, 1 row affected (0.00 sec)

#插入工资
mysql> INSERT INTO payroll VALUES(7,4,'8000','2017-11-15');
Query OK, 1 row affected (0.00 sec)

#插入工资
mysql> INSERT INTO payroll VALUES(8,4,'8500','2017-12-15');
Query OK, 1 row affected (0.00 sec)
```

#插入工资
mysql> INSERT INTO payroll VALUES(9,5,'6000','2017-11-15');
Query OK, 1 row affected (0.00 sec)

#插入工资
mysql> INSERT INTO payroll VALUES(10,5,'6500','2017-12-15');
Query OK, 1 row affected (0.00 sec)

#查看员工表employee中的数据
mysql> SELECT * FROM employee;

id	name	sex	age	deptId
1	张明	男	30	1
2	孙浩	男	25	1
3	张静	女	28	2
4	赵颖	女	32	2
5	刘帅	男	28	2

5 rows in set (0.00 sec)

#查看部门表dept中的数据
mysql> SELECT * FROM dept;

id	deptName	manager
1	软件开发部	王洋
2	人力资源部	吴刚

2 rows in set (0.00 sec)

#查看工资表payroll中的数据
mysql> SELECT * FROM payroll;

id	empId	salary	grantDate
1	1	13500	2017-11-15
2	1	16500	2017-12-15
3	2	9500	2017-11-15
4	2	10500	2017-12-15
5	3	11000	2017-11-15
6	3	12000	2017-12-15
7	4	8000	2017-11-15
8	4	8500	2017-12-15
9	5	6000	2017-11-15
10	5	6500	2017-12-15

10 rows in set (0.00 sec)

#创建视图edp_view,获取员工id、员工姓名、员工部门名称、工资、发放日期
mysql> CREATE VIEW edp_view AS SELECT e.id,e.name,d.deptName,p.salary,p.grantDate
FROM employee e,dept d,payroll p

```
    -> WHERE e.deptId=d.id AND e.id = p.empId;
Query OK, 0 rows affected (0.10 sec)
```

#查看视图表数据
```
mysql> SELECT * FROM edp_view;
+----+------+-----------+--------+------------+
| id | name | deptName  | salary | grantDate  |
+----+------+-----------+--------+------------+
|  1 | 张明 | 软件开发部 |  13500 | 2017-11-15 |
|  1 | 张明 | 软件开发部 |  16500 | 2017-12-15 |
|  2 | 孙浩 | 软件开发部 |   9500 | 2017-11-15 |
|  2 | 孙浩 | 软件开发部 |  10500 | 2017-12-15 |
|  3 | 张静 | 人力资源部 |  11000 | 2017-11-15 |
|  3 | 张静 | 人力资源部 |  12000 | 2017-12-15 |
|  4 | 赵颖 | 人力资源部 |   8000 | 2017-11-15 |
|  4 | 赵颖 | 人力资源部 |   8500 | 2017-12-15 |
|  5 | 刘帅 | 人力资源部 |   6000 | 2017-11-15 |
|  5 | 刘帅 | 人力资源部 |   6500 | 2017-12-15 |
+----+------+-----------+--------+------------+
10 rows in set (0.00 sec)
```

#查看视图里的软件开发部数据
```
mysql> SELECT * FROM edp_view WHERE deptName='软件开发部';
+----+------+-----------+--------+------------+
| id | name | deptName  | salary | grantDate  |
+----+------+-----------+--------+------------+
|  1 | 张明 | 软件开发部 |  13500 | 2017-11-15 |
|  1 | 张明 | 软件开发部 |  16500 | 2017-12-15 |
|  2 | 孙浩 | 软件开发部 |   9500 | 2017-11-15 |
|  2 | 孙浩 | 软件开发部 |  10500 | 2017-12-15 |
+----+------+-----------+--------+------------+
4 rows in set (0.00 sec)
```

#查看视图表里工资超过10000的数据
```
mysql> SELECT * FROM edp_view WHERE salary > 10000;
+----+------+-----------+--------+------------+
| id | name | deptName  | salary | grantDate  |
+----+------+-----------+--------+------------+
|  1 | 张明 | 软件开发部 |  13500 | 2017-11-15 |
|  1 | 张明 | 软件开发部 |  16500 | 2017-12-15 |
|  2 | 孙浩 | 软件开发部 |  10500 | 2017-12-15 |
|  3 | 张静 | 人力资源部 |  11000 | 2017-11-15 |
|  3 | 张静 | 人力资源部 |  12000 | 2017-12-15 |
+----+------+-----------+--------+------------+
5 rows in set (0.00 sec)
```

#查看视图结构
```
mysql> DESC edp_view;
+-------+---------+------+-----+---------+-------+
| Field | Type    | Null | Key | Default | Extra |
+-------+---------+------+-----+---------+-------+
| id    | int(11) | NO   |     | 0       |       |
```

```
| name      | varchar(255) | YES  |     | NULL    |       |
| deptName  | varchar(255) | YES  |     | NULL    |       |
| salary    | varchar(255) | YES  |     | NULL    |       |
| grantDate | date         | YES  |     | NULL    |       |
+-----------+--------------+------+-----+---------+-------+
5 rows in set (0.02 sec)
```

6.1.3 修改视图

对于数据库中已经存在的视图，如果要修改它，有两种方式：第一种方式是使用CREATE OR REPLACE VIEW语句，这种方式可以在视图存在的情况下对视图进行修改，在视图不存在的情况下可创建视图；第二种方式是使用ALTER语句。

使用CREATE OR REPLACE VIEW语句修改视图。

```
mysql> CREATE OR REPLACE VIEW edp_view
    -> AS SELECT e.id,e.name,d.deptName,d.manager,p.salary,p.grantDate
    -> FROM employee e,dept d,payroll p
    -> WHERE e.deptId=d.id AND e.id=p.empId;
```

使用ALTER语句修改视图。

```
mysql> ALTER VIEW edp_view
    -> AS SELECT e.id,e.name,e.sex,d.deptName,d.manager,p.salary,p.grantDate
    -> FROM employee e,dept d,payroll p
    -> WHERE e.deptId=d.id AND e.id=p.empId;
```

实战演练——修改视图

下面在dep_view视图上进行修改，通过使用CREATE OR REPLACE VIEW语句修改视图的方式添加部门经理manager列数据，通过使用ALTER语句修改视图的方式添加性别sex列数据。

```
Microsoft Windows [版本 6.1.7601]
版权所有 (c) 2009 Microsoft Corporation。保留所有权利。

C:\Users\Administrator>mysql –uroot –p123456
mysql: [Warning] Using a password on the command line interface can be insecure.

Welcome to the MySQL monitor.  Commands end with ; or \g.
Your MySQL connection id is 15
Server version: 5.7.20 MySQL Community Server (GPL)

Copyright (c) 2000, 2017, Oracle and/or its affiliates. All rights reserved.

Oracle is a registered trademark of Oracle Corporation and/or its
affiliates. Other names may be trademarks of their respective
owners.

Type 'help;' or '\h' for help. Type '\c' to clear the current input statement.

mysql> USE shop
Database changed

#使用CREATE OR REPLACE VIEW语句修改视图
mysql> CREATE OR REPLACE VIEW edp_view
    -> AS SELECT e.id,e.name,d.deptName,d.manager,p.salary,p.grantDate
```

```
    -> FROM employee e,dept d,payroll p
    -> WHERE e.deptId=d.id AND e.id=p.empId;
Query OK, 0 rows affected (0.09 sec)

#查看视图结构
mysql> DESC edp_view;
+-----------+--------------+------+-----+---------+-------+
| Field     | Type         | Null | Key | Default | Extra |
+-----------+--------------+------+-----+---------+-------+
| id        | int(11)      | NO   |     | 0       |       |
| name      | varchar(255) | YES  |     | NULL    |       |
| deptName  | varchar(255) | YES  |     | NULL    |       |
| manager   | varchar(255) | YES  |     | NULL    |       |
| salary    | varchar(255) | YES  |     | NULL    |       |
| grantDate | date         | YES  |     | NULL    |       |
+-----------+--------------+------+-----+---------+-------+
6 rows in set (0.00 sec)

#查询视图数据
mysql> SELECT * FROM edp_view;
+----+------+-----------+---------+--------+------------+
| id | name | deptName  | manager | salary | grantDate  |
+----+------+-----------+---------+--------+------------+
|  1 | 张明 | 软件开发部 | 王洋    |  13500 | 2017-11-15 |
|  1 | 张明 | 软件开发部 | 王洋    |  16500 | 2017-12-15 |
|  2 | 孙浩 | 软件开发部 | 王洋    |   9500 | 2017-11-15 |
|  2 | 孙浩 | 软件开发部 | 王洋    |  10500 | 2017-12-15 |
|  3 | 张静 | 人力资源部 | 吴刚    |  11000 | 2017-11-15 |
|  3 | 张静 | 人力资源部 | 吴刚    |  12000 | 2017-12-15 |
|  4 | 赵颖 | 人力资源部 | 吴刚    |   8000 | 2017-11-15 |
|  4 | 赵颖 | 人力资源部 | 吴刚    |   8500 | 2017-12-15 |
|  5 | 刘帅 | 人力资源部 | 吴刚    |   6000 | 2017-11-15 |
|  5 | 刘帅 | 人力资源部 | 吴刚    |   6500 | 2017-12-15 |
+----+------+-----------+---------+--------+------------+
10 rows in set (0.00 sec)

#使用ALTER语句修改视图
mysql> ALTER VIEW edp_view
    -> AS SELECT e.id,e.name,e.sex,d.deptName,d.manager,p.salary,p.grantDate
    -> FROM employee e,dept d,payroll p
    -> WHERE e.deptId=d.id AND e.id=p.empId;
Query OK, 0 rows affected (0.11 sec)

#查看视图结构
mysql> DESC edp_view;
+----------+--------------+------+-----+---------+-------+
| Field    | Type         | Null | Key | Default | Extra |
+----------+--------------+------+-----+---------+-------+
| id       | int(11)      | NO   |     | 0       |       |
| name     | varchar(255) | YES  |     | NULL    |       |
| sex      | varchar(10)  | YES  |     | NULL    |       |
| deptName | varchar(255) | YES  |     | NULL    |       |
```

```
| manager   | varchar(255) | YES |     | NULL |     |
| salary    | varchar(255) | YES |     | NULL |     |
| grantDate | date         | YES |     | NULL |     |
+-----------+--------------+-----+-----+------+-----+
7 rows in set (0.00 sec)

#查询视图数据
mysql> SELECT * FROM edp_view;
+----+------+-----+-----------+---------+--------+------------+
| id | name | sex | deptName  | manager | salary | grantDate  |
+----+------+-----+-----------+---------+--------+------------+
|  1 | 张明 | 男  | 软件开发部 | 王洋    | 13500  | 2017-11-15 |
|  1 | 张明 | 男  | 软件开发部 | 王洋    | 16500  | 2017-12-15 |
|  2 | 孙浩 | 男  | 软件开发部 | 王洋    | 9500   | 2017-11-15 |
|  2 | 孙浩 | 男  | 软件开发部 | 王洋    | 10500  | 2017-12-15 |
|  3 | 张静 | 女  | 人力资源部 | 吴刚    | 11000  | 2017-11-15 |
|  3 | 张静 | 女  | 人力资源部 | 吴刚    | 12000  | 2017-12-15 |
|  4 | 赵颖 | 女  | 人力资源部 | 吴刚    | 8000   | 2017-11-15 |
|  4 | 赵颖 | 女  | 人力资源部 | 吴刚    | 8500   | 2017-12-15 |
|  5 | 刘帅 | 男  | 人力资源部 | 吴刚    | 6000   | 2017-11-15 |
|  5 | 刘帅 | 男  | 人力资源部 | 吴刚    | 6500   | 2017-12-15 |
+----+------+-----+-----------+---------+--------+------------+
10 rows in set (0.00 sec)

mysql>
```

6.1.4 更新视图数据

更新视图数据包括修改视图里的数据、插入视图里的数据、删除视图里的数据。基本表可以更新表里的数据，视图也能更新其数据。但是，视图是一张虚拟表，最好用于数据查询，而不要直接更新视图里的数据，应通过基本表来更新数据，然后视图会自动获取新数据。通常，更新视图数据会有很多限制，更新从多个表关联的视图，插入或者删除的时候会失败，不像更新基本表那样方便。

实战演练——更新视图数据

```
Microsoft Windows [版本 6.1.7601]
版权所有 (c) 2009 Microsoft Corporation。保留所有权利。

C:\Users\Administrator>mysql –uroot –p123456
mysql: [Warning] Using a password on the command line interface can be insecure.

Welcome to the MySQL monitor.  Commands end with ; or \g.
Your MySQL connection id is 16
Server version: 5.7.20 MySQL Community Server (GPL)

Copyright (c) 2000, 2017, Oracle and/or its affiliates. All rights reserved.

Oracle is a registered trademark of Oracle Corporation and/or its
affiliates. Other names may be trademarks of their respective
owners.

Type 'help;' or '\h' for help. Type '\c' to clear the current input statement.
```

```
mysql> USE shop
Database changed
```

#查询视图edp_view里的数据
```
mysql> SELECT * FROM edp_view;
+----+------+-----+------------+--------+--------+------------+
| id | name | sex | deptName   | manager| salary | grantDate  |
+----+------+-----+------------+--------+--------+------------+
|  1 | 张明 | 男  | 软件开发部 | 王洋   | 13500  | 2017-11-15 |
|  1 | 张明 | 男  | 软件开发部 | 王洋   | 16500  | 2017-12-15 |
|  2 | 孙浩 | 男  | 软件开发部 | 王洋   |  9500  | 2017-11-15 |
|  2 | 孙浩 | 男  | 软件开发部 | 王洋   | 10500  | 2017-12-15 |
|  3 | 张静 | 女  | 人力资源部 | 吴刚   | 11000  | 2017-11-15 |
|  3 | 张静 | 女  | 人力资源部 | 吴刚   | 12000  | 2017-12-15 |
|  4 | 赵颖 | 女  | 人力资源部 | 吴刚   |  8000  | 2017-11-15 |
|  4 | 赵颖 | 女  | 人力资源部 | 吴刚   |  8500  | 2017-12-15 |
|  5 | 刘帅 | 男  | 人力资源部 | 吴刚   |  6000  | 2017-11-15 |
|  5 | 刘帅 | 男  | 人力资源部 | 吴刚   |  6500  | 2017-12-15 |
+----+------+-----+------------+--------+--------+------------+
10 rows in set (0.01 sec)
```

#更新视图edp_view部门名称为软件开发部的部门经理李姝
```
mysql> UPDATE edp_view SET manager='李姝' WHERE deptName='软件开发部';
Query OK, 1 row affected (0.06 sec)
Rows matched: 1  Changed: 1  Warnings: 0
```

#更新视图里的数据成功，将软件开发部的部门经理改为李姝
```
mysql> SELECT * FROM edp_view;
+----+------+-----+------------+--------+--------+------------+
| id | name | sex | deptName   | manager| salary | grantDate  |
+----+------+-----+------------+--------+--------+------------+
|  1 | 张明 | 男  | 软件开发部 | 李姝   | 13500  | 2017-11-15 |
|  1 | 张明 | 男  | 软件开发部 | 李姝   | 16500  | 2017-12-15 |
|  2 | 孙浩 | 男  | 软件开发部 | 李姝   |  9500  | 2017-11-15 |
|  2 | 孙浩 | 男  | 软件开发部 | 李姝   | 10500  | 2017-12-15 |
|  3 | 张静 | 女  | 人力资源部 | 吴刚   | 11000  | 2017-11-15 |
|  3 | 张静 | 女  | 人力资源部 | 吴刚   | 12000  | 2017-12-15 |
|  4 | 赵颖 | 女  | 人力资源部 | 吴刚   |  8000  | 2017-11-15 |
|  4 | 赵颖 | 女  | 人力资源部 | 吴刚   |  8500  | 2017-12-15 |
|  5 | 刘帅 | 男  | 人力资源部 | 吴刚   |  6000  | 2017-11-15 |
|  5 | 刘帅 | 男  | 人力资源部 | 吴刚   |  6500  | 2017-12-15 |
+----+------+-----+------------+--------+--------+------------+
10 rows in set (0.00 sec)
```

#将部门表dept里的软件开发部的部门经理也改为李姝，联动视图的修改，基本表也进行修改
```
mysql> SELECT * FROM dept;
+----+------------+---------+
| id | deptName   | manager |
+----+------------+---------+
|  1 | 软件开发部 | 李姝    |
|  2 | 人力资源部 | 吴刚    |
```

```
+--+--------+----+
```
2 rows in set (0.00 sec)

#多表关联的视图edp_view,插入数据会失败
mysql> INSERT INTO edp_view VALUES(6,'王鑫','男','办公室','王宇','7000','2017-12-15');
ERROR 1394 (HY000): Can not insert into join view 'shop.edp_view' without fields list

#多表关联的视图edp_view,删除数据会失败
mysql> DELETE edp_view WHERE id=1;
ERROR 1064 (42000): You have an error in your SQL syntax; check the manual that corresponds to your MySQL server version for the right syntax to use near 'WHERE id=1' at line 1

#创建单表的视图emp_view,包含员工id、员工姓名
mysql> CREATE VIEW emp_view
    -> AS SELECT id,name FROM employee;
Query OK, 0 rows affected (0.12 sec)

#查看视图emp_view的结构
mysql> DESC emp_view;
```
+----+-------------+-----+-----+---------+-------+
| Field | Type        | Null | Key | Default | Extra |
+----+-------------+-----+-----+---------+-------+
| id   | int(11)     | NO   |     | 0       |       |
| name | varchar(255)| YES  |     | NULL    |       |
+----+-------------+-----+-----+---------+-------+
```
2 rows in set (0.00 sec)

#查询视图emp_view中的数据
mysql> SELECT * FROM emp_view;
```
+--+------+
| id | name |
+--+------+
| 1 | 张明  |
| 2 | 孙浩  |
| 3 | 张静  |
| 4 | 赵颖  |
| 5 | 刘帅  |
+--+------+
```
5 rows in set (0.00 sec)

#查询员工表employee中的数据
mysql> SELECT * FROM employee;
```
+--+------+-----+-----+------+
| id | name | sex | age | deptId |
+--+------+-----+-----+------+
| 1 | 李浩  | 男  | 30  | 1     |
| 2 | 孙浩  | 男  | 25  | 1     |
| 3 | 张静  | 女  | 28  | 2     |
| 4 | 赵颖  | 女  | 32  | 2     |
| 5 | 刘帅  | 男  | 28  | 2     |
+--+------+-----+-----+------+
```

5 rows in set (0.00 sec)

#在emp_view中插入数据，单表建立的视图插入成功
mysql> INSERT INTO emp_view VALUES(6,'刘涛');
Query OK, 1 row affected (0.00 sec)

#查询视图emp_view中的数据，可以看到插入成功
mysql> SELECT * FROM emp_view;

id	name
1	李浩
2	孙浩
3	张静
4	赵颖
5	刘帅
6	刘涛

6 rows in set (0.00 sec)

#在员工表employee中也插入一条数据
mysql> SELECT * FROM employee;

id	name	sex	age	deptId
1	李浩	男	30	1
2	孙浩	男	25	1
3	张静	女	28	2
4	赵颖	女	32	2
5	刘帅	男	28	2
6	刘涛	NULL	NULL	NULL

6 rows in set (0.00 sec)

#删除视图emp_view，删除成功
mysql> DELETE FROM emp_view WHERE id=3;
Query OK, 1 row affected (0.00 sec)

#查询视图emp_view中的数据
mysql> SELECT * FROM emp_view;

id	name
1	李浩
2	孙浩
4	赵颖
5	刘帅
6	刘涛

5 rows in set (0.00 sec)

#查询员工表employee中的数据

```
mysql> SELECT * FROM employee;
+----+------+------+------+--------+
| id | name | sex  | age  | deptId |
+----+------+------+------+--------+
| 1  | 李浩 | 男   | 30   | 1      |
| 2  | 孙浩 | 男   | 25   | 1      |
| 4  | 赵颖 | 女   | 32   | 2      |
| 5  | 刘帅 | 男   | 28   | 2      |
| 6  | 刘涛 | NULL | NULL | NULL   |
+----+------+------+------+--------+
5 rows in set (0.00 sec)

mysql>
```

6.1.5 删除视图和数据

删除视图就像删除表一样简单，删除视图里的数据要注意两点：如果视图是依赖于单个基本表建立的，则可以直接删除数据；如果视图是依赖于多个基本表建立的，则不允许直接删除数据。

删除视图

DROP VIEW IF EXISTS work_view;
DROP VIEW work_view;

删除视图数据

DELETE FROM work_view;

实战演练——删除视图和数据

Microsoft Windows [版本 6.1.7601]
版权所有 (c) 2009 Microsoft Corporation。保留所有权利。

C:\Users\Administrator>mysql -uroot -p123456
mysql: [Warning] Using a password on the command line interface can be insecure.

Welcome to the MySQL monitor. Commands end with ; or \g.
Your MySQL connection id is 20
Server version: 5.7.20 MySQL Community Server (GPL)

Copyright (c) 2000, 2017, Oracle and/or its affiliates. All rights reserved.

Oracle is a registered trademark of Oracle Corporation and/or its
affiliates. Other names may be trademarks of their respective
owners.

Type 'help;' or '\h' for help. Type '\c' to clear the current input statement.

mysql> USE shop
Database changed

#查看依赖于单个基本表建立的视图emp_view中的数据
mysql> SELECT * FROM emp_view;
+----+------+
| id | name |

```
+--+----+
| 1 | 李浩 |
| 2 | 孙浩 |
| 4 | 赵颖 |
| 5 | 刘帅 |
| 6 | 刘涛 |
+--+----+
5 rows in set (0.00 sec)
```

#依赖于单个基本表建立的视图允许直接删除
mysql> DELETE FROM emp_view WHERE id=5;
Query OK, 1 row affected (0.00 sec)

#查询视图emp_view中的数据
mysql> SELECT * FROM emp_view;
```
+--+-----+
| id | name |
+--+-----+
| 1 | 李浩 |
| 2 | 孙浩 |
| 4 | 赵颖 |
| 6 | 刘涛 |
+--+-----+
4 rows in set (0.00 sec)
```

#查询基本表employee中的数据
mysql> SELECT * FROM employee;
```
+--+----+---+---+-----+
| id | name | sex | age | deptId |
+--+----+---+---+-----+
| 1 | 李浩 | 男 | 30 | 1 |
| 2 | 孙浩 | 男 | 25 | 1 |
| 4 | 赵颖 | 女 | 32 | 2 |
| 6 | 刘涛 | NULL | NULL | NULL |
+--+----+---+---+-----+
4 rows in set (0.00 sec)
```

#如果视图存在，删除视图emp_view
mysql> DROP VIEW IF EXISTS emp_view;
Query OK, 0 rows affected (0.00 sec)

#查看依赖于多个基本表建立的视图edp_view中的数据
mysql> SELECT * FROM edp_view;

id	name	sex	deptName	manager	salary	grantDate
1	李浩	男	软件开发部	李姝	13500	2017-11-15
1	李浩	男	软件开发部	李姝	16500	2017-12-15
2	孙浩	男	软件开发部	李姝	9500	2017-11-15
2	孙浩	男	软件开发部	李姝	10500	2017-12-15
4	赵颖	女	人力资源部	吴刚	8000	2017-11-15
4	赵颖	女	人力资源部	吴刚	8500	2017-12-15

```
+--+----+---+-------+----+----+-----+
6 rows in set (0.00 sec)

#依赖于多个基本表建立的视图是不能直接删除的
mysql> DELETE FROM edp_view WHERE id=1;
ERROR 1395 (HY000): Can not delete from join view 'shop.edp_view'

#删除视图edp_view
mysql> DROP VIEW edp_view;
Query OK, 0 rows affected (0.00 sec)

mysql>
```

6.2 游标

6.2.1 游标的使用

在MySQL数据库查询中，可以使用SELECT语句将满足条件的数据一起查询出来，但是没有办法一行一行地获取数据并做一些复杂的处理。如满足条件的有10条语句，那么使用SELECT语句就一起查询出10条，而获取一行之后接着处理一行数据SELECT语句是做不到的，这时游标就有了用武之地。游标是一个存储在MySQL数据库里用来查询的对象，它可以每次从结果集里获取一行数据，进行相应的处理。有了游标，就可以滚动查询满足条件的数据，可以做一些复杂的处理。MySQL 游标只能用于存储过程或者函数中。

精讲视频

游标

（1）创建游标。使用 DECLARE语句来创建游标，使用FOR关键字来定义相应的 SELECT语句。游标可以在存储过程或者函数中使用，存储过程处理完成后，游标就消失（局限于存储过程）。

```
DECLARE cursor_name  CURSOR FOR selectSql

mysql> DELIMITER $$
mysql> CREATE PROCEDURE p1()              #定义存储过程
    -> BEGIN
    -> DECLARE c_name varchar(20);        #声明变量c_name，用来存放SQL查询出来的姓名name
    -> DECLARE cursor1 CURSOR             #定义游标cursor1
    -> FOR
    -> SELECT name FROM employee;         #SQL语句查询姓名name
    -> OPEN cursor1;                      #打开游标
    -> FETCH cursor1 INTO c_name;         #使用游标，将游标查询出来的姓名name赋值给c_name
    -> SELECT c_name;                     #显示结果
    -> CLOSE cursor1;                     #关闭游标
    -> END
    -> $$
```

（2）打开游标。游标在使用前需要打开，使用OPEN CURSOR语句来打开游标。

```
OPEN cursor1;
```

（3）使用游标。游标被打开后，可以使用FETCH语句访问SQL查询出来的结果集中的每一行数据，它是按顺序进行读取的，不重复读取，执行FETCH语句一次，读取一行数据，如果想读取多行数据，就需要执行FETCH语句多次。

```
FETCH cursor1 INTO c_name
```

（4）关闭游标。游标处理完成后需要关闭，使用CLOSE CURSOR语句来关闭游标。

CLOSE cursor1;

下面创建一个员工表employee，包含员工id、姓名name、工资salary，再创建一个存储过程p2，定义一个游标cursor2，每次输出一行员工id、姓名name、工资salary信息。

实战演练——使用游标

Microsoft Windows [版本 6.1.7601]
版权所有 (c) 2009 Microsoft Corporation。保留所有权利。

C:\Users\Administrator>mysql -uroot -p123456
mysql: [Warning] Using a password on the command line interface can be insecure.

Welcome to the MySQL monitor. Commands end with ; or \g.
Your MySQL connection id is 23
Server version: 5.7.20 MySQL Community Server (GPL)

Copyright (c) 2000, 2017, Oracle and/or its affiliates. All rights reserved.

Oracle is a registered trademark of Oracle Corporation and/or its affiliates. Other names may be trademarks of their respective owners.

Type 'help;' or '\h' for help. Type '\c' to clear the current input statement.

mysql> USE shop
Database changed

#创建员工表employee
mysql> CREATE TABLE employee(
    -> id int not null AUTO_INCREMENT,
    -> name varchar(255),
    -> salary int,
    -> primary key(id)
    ->);
Query OK, 0 rows affected (0.36 sec)

mysql> INSERT INTO employee VALUES(1,'小明',3700);
Query OK, 1 row affected (0.00 sec)

mysql> INSERT INTO employee VALUES(2,'小红',5700);
Query OK, 1 row affected (0.00 sec)

mysql> INSERT INTO employee VALUES(3,'小蓝',7800);
Query OK, 1 row affected (0.00 sec)

#查询员工表数据，使用SELECT语句，一次获取满足条件的所有数据
mysql> SELECT * FROM employee;

```
+----+------+--------+
| id | name | salary |
+----+------+--------+
|  1 | 小明 |   3700 |
|  2 | 小红 |   5700 |
```

```
| 3 | 小蓝  | 7800 |
+---+------+------+
3 rows in set (0.00 sec)
```

#修改结束符$$
```
mysql> DELIMITER $$
```

#创建存储过程p2，定义游标cursor2，读取每一行数据
```
mysql> CREATE PROCEDURE p2()
    -> BEGIN
    -> DECLARE c_id int;
    -> DECLARE c_name varchar(20);
    -> DECLARE c_salary int;
    -> DECLARE cursor2 CURSOR
    -> FOR
    -> SELECT id,name,salary FROM employee;
    -> OPEN cursor2;
    -> FETCH cursor2 INTO c_id,c_name,c_salary;
    -> SELECT c_id,c_name,c_salary;
    -> FETCH cursor2 INTO c_id,c_name,c_salary;
    -> SELECT c_id,c_name,c_salary;
    -> FETCH cursor2 INTO c_id,c_name,c_salary;
    -> SELECT c_id,c_name,c_salary;
    -> CLOSE cursor2;
    -> END
    -> $$
Query OK, 0 rows affected (0.00 sec)
```

#修改结束符
```
mysql> DELIMITER ;
```

#调用存储过程p2
```
mysql> CALL p2();
```

#游标读取一次，输出一次结果
```
+------+--------+----------+
| c_id | c_name | c_salary |
+------+--------+----------+
|    1 | 小明   |     3700 |
+------+--------+----------+
1 row in set (0.00 sec)
```

#游标读取一次，输出一次结果
```
+------+--------+----------+
| c_id | c_name | c_salary |
+------+--------+----------+
|    2 | 小红   |     5700 |
+------+--------+----------+
1 row in set (0.01 sec)
```

#游标读取一次，输出一次结果
```
+------+--------+----------+
```

```
| c_id | c_name | c_salary |
+------+--------+----------+
|    3 | 小蓝   |     7800 |
+------+--------+----------+
1 row in set (0.02 sec)

Query OK, 0 rows affected (0.03 sec)

mysql>
```

6.2.2 游标的WHILE循环

在有多行数据满足条件时,需要逐行来获取数据,而使用FETCH语句每次只能获取一行数据,如果想获取多行数据,就需要将FETCH语句执行多次,在游标里使用循环语句就不需要多次执行FETCH语句。

WHILE循环语句

WHILE……DO……END WHILE

```
WHILE(n > 0) DO
  SELECT n;
  SET n = n - 1;
END WHILE;
```

实战演练——WHILE循环的游标

下面对员工表employee定义存储过程p3,定义游标cursor3,使用WHILE循环语句输出员工id、姓名name、工资salary。

```
Microsoft Windows [版本 6.1.7601]
版权所有 (c) 2009 Microsoft Corporation。保留所有权利。

C:\Users\Administrator>mysql -uroot -p123456
mysql: [Warning] Using a password on the command line interface can be insecure.

Welcome to the MySQL monitor.  Commands end with ; or \g.
Your MySQL connection id is 26
Server version: 5.7.20 MySQL Community Server (GPL)

Copyright (c) 2000, 2017, Oracle and/or its affiliates. All rights reserved.

Oracle is a registered trademark of Oracle Corporation and/or its
affiliates. Other names may be trademarks of their respective
owners.

Type 'help;' or '\h' for help. Type '\c' to clear the current input statement.

mysql> USE shop
Database changed
mysql> SELECT * FROM employee;
+----+------+--------+
| id | name | salary |
+----+------+--------+
|  1 | 小明 |   3700 |
```

```
| 2 | 小红  |  5700 |
| 3 | 小蓝  |  7800 |
+---+------+-------+
3 rows in set (0.06 sec)

mysql> drop procedure p3;
Query OK, 0 rows affected (0.01 sec)

#修改结束符$$
mysql> DELIMITER $$

#定义一个存储过程,先计算出结果总条数,然后使用WHILE进行循环
mysql> CREATE PROCEDURE p3()
    -> BEGIN
    -> DECLARE c_id int;
    -> DECLARE c_name varchar(20);
    -> DECLARE c_salary int;
    -> DECLARE sum int default 0;
    -> DECLARE cursor3 CURSOR
    -> FOR
    -> SELECT id,name,salary FROM employee;
    -> SELECT count(*) INTO sum FROM employee;
    -> OPEN cursor3;
    -> WHILE(sum > 0) DO
    -> FETCH cursor3 INTO c_id,c_name,c_salary,sum;
    -> SELECT c_id,c_name,c_salary;
    -> SET sum = sum -1;
    -> END WHILE;
    -> CLOSE cursor3;
    -> END
    -> $$

#修改结束符
mysql> DELIMITER ;

#调用存储过程,输出每次循环的结果
mysql> call p3();
+------+--------+----------+------+
| c_id | c_name | c_salary | sum  |
+------+--------+----------+------+
|   1  |  小明  |   3700   |  3   |
+------+--------+----------+------+
1 row in set (0.00 sec)

+------+--------+----------+------+
| c_id | c_name | c_salary | sum  |
+------+--------+----------+------+
|   2  |  小红  |   5700   |  2   |
+------+--------+----------+------+
1 row in set (0.02 sec)

+------+--------+----------+------+
```

```
| c_id | c_name | c_salary | sum |
+------+--------+----------+-----+
|   3  | 小蓝   |   7800   |  1  |
+------+--------+----------+-----+
1 row in set (0.03 sec)

Query OK, 0 rows affected (0.04 sec)
```

#删除存储过程p3
mysql>DROP PROCEDURE p3;

#修改结束符$$
mysql> DELIMITER $$

#创建存储过程，使用WHILE循环，并且声明一个变量值，如果循环查不到数据，设置变量结束循环
```
mysql> CREATE PROCEDURE p3()
    -> BEGIN
    -> DECLARE c_id int;
    -> DECLARE c_name varchar(20);
    -> DECLARE c_salary int;
    -> DECLARE flag int default 0;
    -> DECLARE cursor3 CURSOR
    -> FOR
    -> SELECT id,name,salary FROM employee;
    -> DECLARE CONTINUE HANDLER FOR NOT FOUND SET flag = 1;
    -> OPEN cursor3;
    -> WHILE(flag=0) DO
    -> FETCH cursor3 INTO c_id,c_name,c_salary;
    -> SELECT c_id,c_name,c_salary;
    -> END WHILE;
    -> CLOSE cursor3;
    -> END
    -> $$
```

#修改结束符
mysql> DELIMITER ;

#调用存储过程p3，循环输出结果，总条数有3条，却输出4条结果，在WHILE循环中把NOT FOUND变量设置为1时，存储过程还会调用CONTINUE命令继续查询一次，但是调用EXIT命令就不会继续查询
```
mysql> CALL p3();
+------+--------+----------+
| c_id | c_name | c_salary |
+------+--------+----------+
|   1  | 小明   |   3700   |
+------+--------+----------+
1 row in set (0.00 sec)

+------+--------+----------+
| c_id | c_name | c_salary |
+------+--------+----------+
|   2  | 小红   |   5700   |
+------+--------+----------+
```

```
1 row in set (0.01 sec)

+------+--------+----------+
| c_id | c_name | c_salary |
+------+--------+----------+
|    3 | 小蓝   |     7800 |
+------+--------+----------+
1 row in set (0.01 sec)

+------+--------+----------+
| c_id | c_name | c_salary |
+------+--------+----------+
|    3 | 小蓝   |     7800 |
+------+--------+----------+
1 row in set (0.02 sec)
Query OK, 0 rows affected (0.05 sec)
```

#删除存储过程p3
mysql> DROP PROCEDURE p3;

#修改结束符$$
mysql> DELIMITER $$

#使用EXIT来进行变量判断，如果找不到设置变量值为1
mysql> CREATE PROCEDURE p3()
　　-> BEGIN
　　-> DECLARE c_id int;
　　-> DECLARE c_name varchar(20);
　　-> DECLARE c_salary int;
　　-> DECLARE flag int default 0;
　　-> DECLARE cursor3 CURSOR
　　-> FOR
　　-> SELECT id,name,salary FROM employee;
　　-> DECLARE EXIT HANDLER FOR NOT FOUND SET flag = 1;
　　-> OPEN cursor3;
　　-> WHILE(flag=0) DO
　　-> FETCH cursor3 INTO c_id,c_name,c_sal
　　-> SELECT c_id,c_name,c_salary;
　　-> END WHILE;
　　-> CLOSE cursor3;
　　-> END
　　-> $$
Query OK, 0 rows affected (0.00 sec)

#修改结束符
mysql> DELIMITER ;

#调用存储过程
mysql> CALL p3();
```
+------+--------+----------+
| c_id | c_name | c_salary |
+------+--------+----------+
```

```
| 1 | 小明 |  3700 |
+---+------+-------+
1 row in set (0.00 sec)

+------+--------+----------+
| c_id | c_name | c_salary |
+------+--------+----------+
|    2 | 小红   |     5700 |
+------+--------+----------+
1 row in set (0.01 sec)

+------+--------+----------+
| c_id | c_name | c_salary |
+------+--------+----------+
|    3 | 小蓝   |     7800 |
+------+--------+----------+
1 row in set (0.02 sec)

Query OK, 0 rows affected (0.03 sec)

mysql>
```

DECLARE CONTINUE HANDLER FOR NOT FOUND SET flag = 1；该语句用来标志循环如果找不到，就不要继续执行，也可以使用DECLARE CONTINUE HANDLER FOR SQLSTATE '02000' SET flag=0，CONTINUE在找不到后还会执行一次，但是EXIT不会继续执行。

6.2.3 游标的REPEAT循环

游标的循环除了可以使用WHILE循环，也可以使用REPEAT循环，针对员工表employee，定义一个存储过程p4，定义一个游标cursor4，使用REPEAT循环输出结果。

REPEAT循环语句

REPEAT······UNTIL END REPEAT

```
REPEAT
SELECT n;
    SET n = n – 1;
 UNTIL n<0
END REPEAT
```

实战演练——REPEAT循环的游标

Microsoft Windows [版本 6.1.7601]
版权所有 (c) 2009 Microsoft Corporation。保留所有权利。

C:\Users\Administrator>mysql –uroot –p123456
mysql: [Warning] Using a password on the command line interface can be insecure.

Welcome to the MySQL monitor. Commands end with ; or \g.
Your MySQL connection id is 27
Server version: 5.7.20 MySQL Community Server (GPL)

Copyright (c) 2000, 2017, Oracle and/or its affiliates. All rights reserved.

Oracle is a registered trademark of Oracle Corporation and/or its affiliates. Other names may be trademarks of their respective owners.

Type 'help;' or '\h' for help. Type '\c' to clear the current input statement.

mysql> USE shop
Database changed

#查询员工表employee
mysql> SELECT * FROM employee;

```
+----+------+--------+
| id | name | salary |
+----+------+--------+
|  1 | 小明 |   3700 |
|  2 | 小红 |   5700 |
|  3 | 小蓝 |   7800 |
+----+------+--------+
3 rows in set (0.00 sec)
```

#修改结束符$$
mysql> DELIMITER $$

#定义存储过程p4，定义游标cursor4，使用REPEAT循环
mysql> CREATE PROCEDURE p4()
　　-> BEGIN
　　-> DECLARE c_id int;
　　-> DECLARE c_name varchar(20);
　　-> DECLARE c_salary int;
　　-> DECLARE flag int default 0;
　　-> DECLARE cursor4 CURSOR
　　-> FOR
　　-> SELECT id,name,salary FROM employee;
　　-> DECLARE EXIT HANDLER FOR NOT FOUND SET flag = 1;
　　-> OPEN cursor4;
　　-> REPEAT
　　-> FETCH cursor4 INTO c_id,c_name,c_salary;
　　-> SELECT c_id,c_name,c_salary;
　　-> UNTIL flag=1
　　-> END REPEAT;
　　-> CLOSE cursor4;
　　-> END
　　-> $$
Query OK, 0 rows affected (0.00 sec)

#修改结束符
mysql> DELIMITER ;

#调用存储过程p4，循环输出结果
mysql> CALL p4();

```
+------+--------+----------+
| c_id | c_name | c_salary |
+------+--------+----------+
|    1 | 小明   |     3700 |
+------+--------+----------+
1 row in set (0.00 sec)

+------+--------+----------+
| c_id | c_name | c_salary |
+------+--------+----------+
|    2 | 小红   |     5700 |
+------+--------+----------+
1 row in set (0.01 sec)

+------+--------+----------+
| c_id | c_name | c_salary |
+------+--------+----------+
|    3 | 小蓝   |     7800 |
+------+--------+----------+
1 row in set (0.03 sec)

Query OK, 0 rows affected (0.04 sec)

mysql>
```

6.2.4 游标的LOOP循环

游标的循环除了使用WHILE、REPEAT循环，也可以使用LOOP循环。下面针对员工表employee，定义一个存储过程p5，定义一个游标cursor5，使用LOOP循环输出结果。

LOOP循环语句

LOOP……END LOOP

```
DECLARE i int default 1;
lp1：LOOP
SET i = i+1;
IF i > 30 THEN
LEAVE lp1；#离开循环体
END IF;
END LOOP;
```

实战演练——LOOP循环的游标

```
Microsoft Windows [版本 6.1.7601]
版权所有 (c) 2009 Microsoft Corporation。保留所有权利。

C:\Users\Administrator>mysql –uroot –p123456
mysql: [Warning] Using a password on the command line interface can be insecure.

Welcome to the MySQL monitor.  Commands end with ; or \g.
Your MySQL connection id is 28
Server version: 5.7.20 MySQL Community Server (GPL)

Copyright (c) 2000, 2017, Oracle and/or its affiliates. All rights reserved.
```

```
Oracle is a registered trademark of Oracle Corporation and/or its
affiliates. Other names may be trademarks of their respective
owners.

Type 'help;' or '\h' for help. Type '\c' to clear the current input statement.

mysql> USE shop
Database changed

#查看员工表employee中的数据
mysql> SELECT * FROM employee;
+----+------+--------+
| id | name | salary |
+----+------+--------+
|  1 | 小明 |   3700 |
|  2 | 小红 |   5700 |
|  3 | 小蓝 |   7800 |
+----+------+--------+
3 rows in set (0.00 sec)

#修改结束符$$
mysql> DELIMITER $$

#创建存储过程p5，定义游标cursor5，使用LOOP循环
mysql> CREATE PROCEDURE p5()
    -> BEGIN
    -> DECLARE c_id int;
    -> DECLARE c_name varchar(20);
    -> DECLARE c_salary int;
    -> DECLARE flag int default 0;
    -> DECLARE cursor5 CURSOR
    -> FOR
    -> SELECT id,name,salary FROM employee;
    -> DECLARE EXIT HANDLER FOR NOT FOUND SET flag = 1;
    -> OPEN cursor5;
    -> loop_label: LOOP    #循环开始
    ->     FETCH cursor5 INTO c_id,c_name,c_salary;
    ->     SELECT c_id,c_name,c_salary;
    ->     IF(flag=1)THEN
    ->        LEAVE  loop_label;   #终止循环
    ->     END IF;
    ->   END LOOP;
    -> CLOSE cursor5;
    -> END
    -> $$
Query OK, 0 rows affected (0.00 sec)

#修改结束符
mysql> DELIMITER ;

#调用存储过程p5
```

```
mysql> CALL p5();
+------+--------+----------+
| c_id | c_name | c_salary |
+------+--------+----------+
|    1 | 小明   |     3700 |
+------+--------+----------+
1 row in set (0.00 sec)

+------+--------+----------+
| c_id | c_name | c_salary |
+------+--------+----------+
|    2 | 小红   |     5700 |
+------+--------+----------+
1 row in set (0.02 sec)

+------+--------+----------+
| c_id | c_name | c_salary |
+------+--------+----------+
|    3 | 小蓝   |     7800 |
+------+--------+----------+
1 row in set (0.03 sec)

Query OK, 0 rows affected (0.04 sec)

mysql>
```

6.3 触发器

触发器是一种与表有关的操作对象,当表上出现指定事件(INSERT、UPDATE、DELETE)时,会调用触发器对象,执行触发器的操作,如在主表上执行删除语句,可以调用触发器执行删除子表语句,以达到级联删除的效果。

6.3.1 创建触发器

创建触发器和创建存储过程类似,基本语法如下。

```
CREATE
    [DEFINER = { user | CURRENT_USER }]
    TRIGGER trigger_name
    trigger_time trigger_event
    ON tbl_name FOR EACH ROW
    trigger_body

CREATE
    TRIGGER add_data
    AFTER INSERT
    ON t1 FOR EACH ROW
    BEGIN
      INSERT INTO t2 VALUES('你好');
END
```

(1)TRIGGER是触发器的关键字,用来标识触发器。

(2)trigger_name:触发器名称,不能与已有触发器的名称重复。

（3）trigger_time：触发器触发的时机，只有两个值，即BEFORE和AFTER。
（4）trigger_event：触发器触发的事件，只有三个值，即INSERT、UPDATE、DELETE。
（5）tbl_name：建立触发器的表名，就是在哪张表添加触发器。
（6）FOR EACH ROW：在表的每一行操作。
（7）trigger_body：触发器程序体，可以是一句SQL语句，或者用 BEGIN 和 END 包含的多条语句。
（8）根据触发的时机和事件，可以建立6种类型的触发器：插入之前（BEFORE INSERT）触发器、更新之前（BEFORE UPDATE）触发器、删除之前（BEFORE DELETE）触发器、插入之后（AFTER INSERT）触发器、更新之后（AFTER UPDATE）触发器、删除之后（AFTER DELETE）触发器。
（9）一个表上不能同时建立两个类型相同的触发器，如建立一个插入之前触发器后，再建立一个插入之前触发器是不允许的。

实战演练——创建触发器

下面创建一个用户表user，包含主键id、用户名字name、性别sex三个字段，再创建一个统计表stat，包含主键id、数量num两个字段，再创建一个触发器，user表中每插入一条数据，统计表stat中的数量num加1。

```
Microsoft Windows [版本 6.1.7601]
版权所有 (c) 2009 Microsoft Corporation。保留所有权利。

C:\Users\Administrator>mysql -uroot -p123456
mysql: [Warning] Using a password on the command line interface can be insecure.

Welcome to the MySQL monitor.  Commands end with ; or \g.
Your MySQL connection id is 22
Server version: 5.7.20 MySQL Community Server (GPL)

Copyright (c) 2000, 2017, Oracle and/or its affiliates. All rights reserved.

Oracle is a registered trademark of Oracle Corporation and/or its
affiliates. Other names may be trademarks of their respective
owners.

Type 'help;' or '\h' for help. Type '\c' to clear the current input statement.

mysql> use shop
Database changed

#创建用户表user
mysql> CREATE TABLE user(
    -> id int not null AUTO_INCREMENT,
    -> name varchar(255),
    -> sex varchar(10),
    -> primary key(id)
    -> );
Query OK, 0 rows affected (1.23 sec)

#创建统计表stat
mysql> CREATE TABLE stat(
    -> id int not null AUTO_INCREMENT,
    -> num int,
```

```
        -> primary key(id)
        -> );
Query OK, 0 rows affected (0.34 sec)
```

#在统计表中插入一条数据，默认值为0
```
mysql> INSERT INTO stat VALUES (1,0);
Query OK, 1 row affected (0.00 sec)
```

#修改结束符$$
```
mysql> DELIMITER $$
```

#创建一个触发器来计算用户的数量，统计数量的值放置在stat的num字段里
```
mysql> CREATE TRIGGER cpuNum
    -> AFTER INSERT
    -> ON user FOR EACH ROW
    -> BEGIN
    -> UPDATE stat set num = num +1 WHERE id = 1;
    -> END
    -> $$
Query OK, 0 rows affected (0.18 sec)
```

#修改结束符
```
mysql> DELIMITER ;
```

#查看统计表stat中的数量num=0
```
mysql> SELECT * FROM stat;
+----+-----+
| id | num |
+----+-----+
|  1 |  0  |
+----+-----+
1 row in set (0.00 sec)
```

#在用户表user中插入一条数据
```
mysql> INSERT INTO user VALUES(1,'小明','男');
Query OK, 1 row affected (0.01 sec)
```

#统计表stat中的num加1
```
mysql> SELECT * FROM stat;
+----+-----+
| id | num |
+----+-----+
|  1 |  1  |
+----+-----+
1 row in set (0.00 sec)
```

#在用户表user中再插入一条数据
```
mysql> INSERT INTO user VALUES(2,'小红','女');
Query OK, 1 row affected (0.00 sec)
```

#统计表stat中的num再加1
```
mysql> SELECT * FROM stat;
```

```
+----+-----+
| id | num |
+----+-----+
|  1 |  2  |
+----+-----+
1 row in set (0.00 sec)

mysql>
```

6.3.2 NEW和OLD关键字

触发器用于一张表的某一行数据，如果想在触发器里使用到这行数据，怎么办呢？MySQL数据库定义了NEW和OLD关键字，用来记录触发器的那一行数据。

（1）对于INSERT型触发器，NEW关键字用来表示将要（BEFORE）或已经（AFTER）插入的新数据。

（2）对于UPDATE型触发器，OLD关键字用来表示将要（BEFORE）或已经（AFTER）被修改的原数据，NEW用来表示将要（BEFORE）或已经（AFTER）修改为的新数据。

（3）对于DELETE 型触发器，OLD 用来表示将要（BEFORE）或已经（AFTER）被删除的原数据。

（4）OLD 是只读的，而 NEW 则可以在触发器中使用 SET 赋值。

（5）使用方法：NEW.columnName（columnName 列名）。

实战演练 ——使用NEW和OLD关键字

下面创建一个员工表employee，包含字段主键员工id、姓名name、性别sex、年龄age、部门外键deptId；创建一个部门表dept，包含字段主键部门id、部门名称deptName、部门经理manager；在部门表上添加一个触发器，在删除部门后，触发器会删除部门下的员工。

```
Microsoft Windows [版本 6.1.7601]
版权所有 (c) 2009 Microsoft Corporation。保留所有权利。

C:\Users\Administrator>mysql -uroot -p123456
mysql: [Warning] Using a password on the command line interface can be insecure.

Welcome to the MySQL monitor.  Commands end with ; or \g.
Your MySQL connection id is 12
Server version: 5.7.20 MySQL Community Server (GPL)

Copyright (c) 2000, 2017, Oracle and/or its affiliates. All rights reserved.

Oracle is a registered trademark of Oracle Corporation and/or its
affiliates. Other names may be trademarks of their respective
owners.

Type 'help;' or '\h' for help. Type '\c' to clear the current input statement.

mysql> USE shop
Database changed

#创建员工表employee
mysql> CREATE TABLE employee(
    -> id int not null AUTO_INCREMENT,
    -> name varchar(255),
```

```
    -> sex varchar(10),
    -> age int,
    -> deptId int,
    -> primary key(id)
    -> );
Query OK, 0 rows affected (0.34 sec)
```

#创建部门表dept
```
mysql> CREATE TABLE dept(
    -> id int not null AUTO_INCREMENT,
    -> deptName varchar(255),
    -> manager varchar(255),
    -> primary key(id)
    -> );
Query OK, 0 rows affected (0.38 sec)
```

#创建工资表payroll
```
mysql> CREATE TABLE payroll(
    -> id int not null AUTO_INCREMENT,
    -> empId int,
    -> salary varchar(255),
    -> grantDate date,
    -> primary key(id)
    -> );
Query OK, 0 rows affected (0.46 sec)
```

#插入员工
```
mysql> INSERT INTO employee VALUES(1,'张明','男',30,1);
Query OK, 1 row affected (0.08 sec)
```

#插入员工
```
mysql> INSERT INTO employee VALUES(2,'孙浩','男',25,1);
Query OK, 1 row affected (0.00 sec)
```

#插入员工
```
mysql> INSERT INTO employee VALUES(3,'张静','女',28,2);
Query OK, 1 row affected (0.00 sec)
```

#插入员工
```
mysql> INSERT INTO employee VALUES(4,'赵颖','女',32,2);
Query OK, 1 row affected (0.00 sec)
```

#插入员工
```
mysql> INSERT INTO employee VALUES(5,'刘帅','男',28,2);
Query OK, 1 row affected (0.00 sec)
```

#插入部门
```
mysql> INSERT INTO dept VALUES(1,'软件开发部','王洋');
Query OK, 1 row affected (0.00 sec)
```

#插入部门
```
mysql> INSERT INTO dept VALUES(2,'人力资源部','吴刚');
```

Query OK, 1 row affected (0.00 sec)

#查询员工表中的数据
```
mysql> SELECT * FROM employee;
+----+------+-----+-----+--------+
| id | name | sex | age | deptId |
+----+------+-----+-----+--------+
|  1 | 李浩 | 男  |  30 |      1 |
|  2 | 孙浩 | 男  |  25 |      1 |
|  3 | 张静 | 女  |  28 |      2 |
|  4 | 赵颖 | 女  |  32 |      2 |
|  5 | 刘帅 | 男  |  28 |      2 |
+----+------+-----+-----+--------+
5 rows in set (0.00 sec)
```

#查询部门表中的数据
```
mysql> SELECT * FROM dept;
+----+------------+---------+
| id | deptName   | manager |
+----+------------+---------+
|  1 | 软件开发部 | 王洋    |
|  2 | 人力资源部 | 吴刚    |
+----+------------+---------+
2 rows in set (0.00 sec)
```

#修改结束符为$$
```
mysql> DELIMITER $$
```

#创建按部门删除员工的触发器，删除部门后，触发器会自动删除这个部门的员工
```
mysql> CREATE TRIGGER deleteEmp
    -> AFTER DELETE
    -> ON dept FOR EACH ROW
    -> BEGIN
    -> DELETE FROM employee WHERE deptId = OLD.id;
    -> END
    -> $$
Query OK, 0 rows affected (0.21 sec)
```

#修改结束符为$$
```
mysql> DELIMITER ;
```
#删除部门id等于1，触发器会删除这个部门下的员工
```
mysql> DELETE FROM dept WHERE id = 1;
Query OK, 1 row affected (0.08 sec)
```

#查询部门表中的数据
```
mysql> SELECT * FROM dept;
+----+------------+---------+
| id | deptName   | manager |
+----+------------+---------+
|  2 | 人力资源部 | 吴刚    |
+----+------------+---------+
1 row in set (0.00 sec)
```

#查询员工表中的数据，可以看到deptId等于1的员工被删除
```
mysql> SELECT * FROM employee;
+----+------+-----+-----+--------+
| id | name | sex | age | deptId |
+----+------+-----+-----+--------+
|  3 | 张静 | 女  | 28  |   2    |
|  4 | 赵颖 | 女  | 32  |   2    |
|  5 | 刘帅 | 男  | 28  |   2    |
+----+------+-----+-----+--------+
3 rows in set (0.00 sec)

mysql>
```

6.3.3 查看和删除触发器

（1）查看触发器。查看触发器和查看数据库表一样，通过SHOW TRIGGERS就可以查看触发器；

```
mysql> SHOW TRIGGERS \G
*************************** 1. row ***************************
             Trigger: deleteEmp                   #触发器名称
               Event: DELETE                      #触发器事件
               Table: dept                        #触发的表
           Statement: BEGIN                       #执行的SQL语句
                        DELETE FROM employee WHERE deptId = OLD.id;
                     END
              Timing: AFTER                       #执行的时机
             Created: 2018-01-06 21:45:04.23      #创建时间
            sql_mode: STRICT_TRANS_TABLES,NO_ENGINE_SUBSTITUTION
             Definer: root@localhost   #定义的用户
character_set_client: utf8             #编码
collation_connection: utf8_general_ci
  Database Collation: utf8_general_ci
```

（2）删除触发器。删除触发器就像删除数据库表一样简单，使用关键字TRIGGER来标识触发器，然后使用DROP操作删除指定名称的触发器。

DROP TRIGGER trigger_name；

mysql> DROP TRIGGER cpuNum；

6.3.4 INSERT型触发器

INSERT型触发器分为插入之前（BEFORE）触发器和插入之后（AFTER）触发器，NEW关键字用来表示将要（BEFORE）或已经（AFTER）插入的新数据。

下面创建一个员工表employee，包含员工id、姓名name、工资salary，再创建一个记录表record，包含员工id、姓名name、工资salary，如果员工工资低于3 500元，则记录到低工资record表中，否则就不记录。

实战演练——INSERT型触发器

Microsoft Windows [版本 6.1.7601]
版权所有 (c) 2009 Microsoft Corporation。保留所有权利。

C:\Users\Administrator>mysql -uroot -p123456
mysql: [Warning] Using a password on the command line interface can be insecure.

```
Welcome to the MySQL monitor.  Commands end with ; or \g.
Your MySQL connection id is 23
Server version: 5.7.20 MySQL Community Server (GPL)

Copyright (c) 2000, 2017, Oracle and/or its affiliates. All rights reserved.

Oracle is a registered trademark of Oracle Corporation and/or its
affiliates. Other names may be trademarks of their respective
owners.

Type 'help;' or '\h' for help. Type '\c' to clear the current input statement.

mysql> USE shop
Database changed

#创建员工表employee
mysql> CREATE TABLE employee(
    -> id int not null AUTO_INCREMENT,
    -> name varchar(255),
    -> salary int,
    -> primary key(id)
    -> );
Query OK, 0 rows affected (0.36 sec)

#创建记录表，来记录工资低于3 500元的员工
mysql> CREATE TABLE record(
    -> id int not null AUTO_INCREMENT,
    -> name varchar(255),
    -> salary int,
    -> primary key(id)
    -> );
Query OK, 0 rows affected (0.32 sec)

#查询员工表employee中的数据，目前没有数据
mysql> SELECT * FROM employee;
Empty set (0.00 sec)

#查询记录表record中的数据，目前没有数据
mysql> SELECT * FROM record;
Empty set (0.00 sec)

#修改结束符$$
mysql> DELIMITER $$

#创建触发器，添加低于3 500元的工资到记录表里
mysql> CREATE TRIGGER addRecord
    -> AFTER INSERT
    -> ON employee FOR EACH ROW
    -> BEGIN
    -> IF(NEW.salary <3500) THEN
    -> INSERT INTO record VALUES(NEW.id,NEW.name,NEW.salary);
```

```
    -> END IF;
    -> END
    -> $$
Query OK, 0 rows affected (0.17 sec)
```

#修改结束符
mysql> DELIMITER ;

#在员工表里插入一条工资低于3 500元的员工记录
mysql> INSERT INTO employee VALUES(1,'小明',2800);
```
Query OK, 1 row affected (0.00 sec)
```

#查询员工表employee中的数据，可以看到插入成功
mysql> SELECT * FROM employee;
```
+----+------+--------+
| id | name | salary |
+----+------+--------+
|  1 | 小明 |   2800 |
+----+------+--------+
1 row in set (0.00 sec)
```

#查询记录表record中的数据，可以看到插入一条低于3 500元工资的员工记录
mysql> SELECT * FROM record;
```
+----+------+--------+
| id | name | salary |
+----+------+--------+
|  1 | 小明 |   2800 |
+----+------+--------+
1 row in set (0.00 sec)
```

#再插入一条高于3 500元工资的员工记录
mysql> INSERT INTO employee VALUES(2,'小红',5600);
```
Query OK, 1 row affected (0.00 sec)
```

#查询员工表employee中的数据，可以看到插入成功
mysql> SELECT * FROM employee;
```
+----+------+--------+
| id | name | salary |
+----+------+--------+
|  1 | 小明 |   2800 |
|  2 | 小红 |   5600 |
+----+------+--------+
2 rows in set (0.00 sec)
```

#查询记录表record中的数据，工资高于3 500元的员工没有插入
mysql> SELECT * FROM record;
```
+----+------+--------+
| id | name | salary |
+----+------+--------+
|  1 | 小明 |   2800 |
+----+------+--------+
1 row in set (0.00 sec)
```

mysql>

6.3.5 UPDATE型触发器

UPDATE型触发器分为更新之前（BEFORE）触发器和更新之后（AFTER）触发器。对于UPDATE型触发器，OLD关键字用来表示将要（BEFORE）或已经（AFTER）被修改的原数据，NEW用来表示将要（BEFORE）或已经（AFTER）修改为的新数据。

在插入型触发器里已经建好了两个表：员工表employee、记录表record，工资低于3 500元的员工会被记录到record表里，如果员工涨工资后工资超过了3 500元，那么就得从记录表record中删除这名员工，如果员工更改工资后工资低于3 500元，那么就要更改记录表record中的数据，这时就需要使用UPDATE型触发器。

实战演练 ——UPDATE型触发器

Microsoft Windows [版本 6.1.7601]
版权所有 (c) 2009 Microsoft Corporation。保留所有权利。

C:\Users\Administrator>mysql –uroot –p123456
mysql: [Warning] Using a password on the command line interface can be insecure.

Welcome to the MySQL monitor. Commands end with ; or \g.
Your MySQL connection id is 24
Server version: 5.7.20 MySQL Community Server (GPL)

Copyright (c) 2000, 2017, Oracle and/or its affiliates. All rights reserved.

Oracle is a registered trademark of Oracle Corporation and/or its
affiliates. Other names may be trademarks of their respective
owners.

Type 'help;' or '\h' for help. Type '\c' to clear the current input statement.

mysql> USE shop
Database changed

#查看员工表employee中的数据
mysql> SELECT * FROM employee;
+----+------+--------+
| id | name | salary |
+----+------+--------+
| 1 | 小明 | 2800 |
| 2 | 小红 | 5600 |
+----+------+--------+
2 rows in set (0.00 sec)

#查看记录表中的数据
mysql> SELECT * FROM record;
+----+------+--------+
| id | name | salary |
+----+------+--------+
| 1 | 小明 | 2800 |
+----+------+--------+

1 row in set (0.00 sec)

#修改结束符
mysql> DELIMITER $$

#创建更新触发器，在employee工资表中进行修改，工资大于3 500元，则删除记录表中的对应数据，否则插入或者修改记录表中的数据，REPLACE可以用于插入或者更新记录

```
mysql> CREATE TRIGGER updateRecord
    -> AFTER UPDATE
    -> ON employee FOR EACH ROW
    -> BEGIN
    -> IF(NEW.salary > 3500) THEN
    -> DELETE FROM record WHERE id=NEW.id;
    -> ELSE
    -> REPLACE INTO record VALUES(NEW.id,NEW.name,NEW.salary);
    -> END IF;
    -> END
    -> $$
Query OK, 0 rows affected (0.16 sec)
```

#修改结束符
mysql> DELIMITER ;

#将员工id为1的工资从2 800元更改为6 000元，在记录表中应删除员工id为1的记录
mysql> UPDATE employee SET salary = 6000 WHERE id=1;
Query OK, 1 row affected (0.00 sec)
Rows matched: 1 Changed: 1 Warnings: 0

#修改成功
mysql> SELECT * FROM employee;
+----+------+--------+
| id | name | salary |
+----+------+--------+
| 1 | 小明 | 6000 |
| 2 | 小红 | 5600 |
+----+------+--------+
2 rows in set (0.00 sec)

#工资大于3 500元，在记录表中删除员工id为1的数据
mysql> SELECT * FROM record;
Empty set (0.00 sec)

#将员工id为2的工资从5 600元改为2 000元，在记录表中应插入记录
mysql> UPDATE employee SET salary = 2000 WHERE id=2;
Query OK, 1 row affected (0.00 sec)
Rows matched: 1 Changed: 1 Warnings: 0

#修改成功
mysql> SELECT * FROM employee;
+----+------+--------+
| id | name | salary |
+----+------+--------+

```
| 1 | 小明 |  6000 |
| 2 | 小红 |  2000 |
+----+------+-------+
2 rows in set (0.00 sec)
```

#插入员工id为2的记录，工资小于3 500元
mysql> SELECT * FROM record;
```
+----+------+--------+
| id | name | salary |
+----+------+--------+
|  2 | 小红 |  2000  |
+----+------+--------+
1 row in set (0.00 sec)
```

#将员工id为2的工资从2 000元更改为2 500元，工资还是低于3 500元，应该更改记录表中的数据
mysql> UPDATE employee SET salary = 2500 WHERE id=2;
Query OK, 1 row affected (0.06 sec)
Rows matched: 1 Changed: 1 Warnings: 0

#修改成功
mysql> SELECT * FROM employee;
```
+----+------+--------+
| id | name | salary |
+----+------+--------+
|  1 | 小明 |  6000  |
|  2 | 小红 |  2500  |
+----+------+--------+
2 rows in set (0.00 sec)
```

#修改成功
mysql> SELECT * FROM record;
```
+----+------+--------+
| id | name | salary |
+----+------+--------+
|  2 | 小红 |  2500  |
+----+------+--------+
1 row in set (0.00 sec)
```

mysql>

6.3.6 DELETE型触发器

DELETE型触发器分为删除之前（BEFORE）触发器和删除之后（AFTER）触发器，对于DELETE型触发器，OLD用来表示将要（BEFORE）或已经（AFTER）被删除的原数据。

在插入型触发器里已经建好了两个表：员工表employee、记录表record，工资低于3 500元的员工会被记录到record表里。如果在员工表中进行了删除员工的操作，那么在记录表中若记录了该员工，则应该一起删除，这时就需要使用DELETE型触发器。

实战演练——DELETE型触发器

Microsoft Windows [版本 6.1.7601]
版权所有 (c) 2009 Microsoft Corporation。保留所有权利。

```
C:\Users\Administrator>mysql -uroot -p123456
mysql: [Warning] Using a password on the command line interface can be insecure.

Welcome to the MySQL monitor.  Commands end with ; or \g.
Your MySQL connection id is 25
Server version: 5.7.20 MySQL Community Server (GPL)

Copyright (c) 2000, 2017, Oracle and/or its affiliates. All rights reserved.

Oracle is a registered trademark of Oracle Corporation and/or its
affiliates. Other names may be trademarks of their respective
owners.

Type 'help;' or '\h' for help. Type '\c' to clear the current input statement.

mysql> USE shop
Database changed

#出现员工表employee中的数据
mysql> SELECT * FROM employee;
+----+------+------+
| id | name | salary |
+----+------+------+
|  1 | 小明 | 6000 |
|  2 | 小红 | 2500 |
+----+------+------+
2 rows in set (0.00 sec)

#查询记录表record中的数据
mysql> SELECT * FROM record;
+----+------+------+
| id | name | salary |
+----+------+------+
|  2 | 小红 | 2500 |
+----+------+------+
1 row in set (0.00 sec)

#修改结束符$$
mysql> DELIMITER $$

#创建删除触发器，删除记录表中的数据
mysql> CREATE TRIGGER delRecord
    -> AFTER DELETE
    -> ON employee FOR EACH ROW
    -> BEGIN
    -> DELETE FROM record WHERE id = OLD.id;
    -> END
    -> $$
Query OK, 0 rows affected (0.17 sec)

#修改结束符
```

```
mysql> DELIMITER ;

#删除员工id为2的记录
mysql> DELETE FROM employee WHERE id = 2;
Query OK, 1 row affected (0.00 sec)

#查询员工表employee中的数据，员工id为2的数据已被删除
mysql> SELECT * FROM employee;
+----+------+--------+
| id | name | salary |
+----+------+--------+
|  1 | 小明 |  6000  |
+----+------+--------+
1 row in set (0.00 sec)

#记录表中员工id为2的记录也被删除
mysql> SELECT * FROM record;
Empty set (0.00 sec)

#删除员工id为1的记录
mysql> DELETE FROM employee WHERE id = 1;
Query OK, 1 row affected (0.00 sec)

#删除成功
mysql> SELECT * FROM employee;
Empty set (0.00 sec)

#没有记录，则不进行删除
mysql> SELECT * FROM record;
Empty set (0.00 sec)

mysql>
```

6.4 小结

本章讲解了MySQL高级特性，包括视图、游标和触发器。视图要学会什么是视图、如何创建视图、如何修改视图、如何更新视图、如何删除视图和数据；游标要学会游标的使用、在游标中学会WHILE、REPEAT、LOOP循环的使用；触发器要学会如何创建触发器、查看触发器、删除触发器以及INSERT型触发器的使用、UPDATE型触发器的使用、DELETE型触发器的使用。掌握了这些高级特性，我们对于数据库的应用和编程就会得心应手，也能更高效地应用MySQL数据库。

第7章
索引

本章要点

- 索引的基本语法
- 常见的查询算法
- 索引的数据结构
- 索引实现原理
- 索引的应用
- 索引的类型
- 索引不能使用
- 索引的利弊及建立原则
- 小结

■ 索引是用来提高数据库查询效率的一种方式。建立有效、合适的索引可以极大地提高查询效率,是数据库高性能优化操作必做的一件事。索引通俗地讲就像书籍的目录一样,它对数据库表也维护一套目录,通过目录可以快速地定位到要查询的数据,而不用把整个数据库表都扫描一遍,所以索引是提高SQL语句查询效率的一种有效手段。

7.1 索引的基本语法

7.1.1 创建索引

创建索引有三种方式：第一种是在创建表的时候创建索引；第二种是使用CREATE INDEX创建索引；第三种是使用ALTER TABLE创建索引。

（1）创建表的时候创建索引。

精讲视频

索引的基本语法

```
CREATE TABLE  表名( 属性名 数据类型[完整性约束条件],
属性名 数据类型[完整性约束条件],
......
属性名 数据类型
[ UNIQUE | FULLTEXT | SPATIAL ] INDEX | KEY
[ 别名 ] ( 属性名1 [(长度)] [ ASC | DESC ] )
);

mysql> CREATE TABLE student(
    -> id int primary key,
    -> name varchar(255),
    -> sex varchar(10),
    -> index(name)
    -> );
```

① UNIQUE：代表索引是唯一索引。
② FULLTEXT：代表索引是全文索引。
③ SPATIAL：代表索引是空间索引。
④ INDEX和KEY：用来指定某个字段为索引，使用效果一样。
⑤ 别名：用来给创建的索引取新名称。
⑥ 属性1：用来指定索引对应的字段的名称，该字段必须为前面定义好的字段。
⑦ 长度：指索引的长度，必须是字符串类型才可以使用。
⑧ ASC：表示升序排列。
⑨ DESC：表示降序排列。

（2）使用CREATE INDEX创建索引。

```
CREATE INDEX index_name ON table_name (column_list)
CREATE UNIQUE INDEX index_name ON table_name (column_list)

mysql> CREATE INDEX sex_index ON student(sex);
```

table_name是要增加索引的表名，column_list代表哪些列要加索引，有多列时各列之间用逗号分隔，索引名index_name可选，如果没有写index_name将根据第一个索引列赋一个名称。

CREATE INDEX可对表增加普通索引或UNIQUE唯一索引，不能用CREATE INDEX语句创建PRIMARY KEY主键索引。

（3）使用ALTER TABLE创建索引。

```
ALTER TABLE table_name ADD INDEX index_name (column_list)
ALTER TABLE table_name ADD UNIQUE (column_list)
ALTER TABLE table_name ADD PRIMARY KEY (column_list)

mysql> ALTER TABLE student ADD INDEX name_index(name);
```

ALTER TABLE用来创建普通索引、UNIQUE索引或PRIMARY KEY索引。

7.1.2 查看索引

```
SHOW INDEX FROM table_name;
SHOW KEYS FROM table_name;
```

```
mysql> SHOW INDEX FROM student;
+---------+------------+--------------+--------------+-------------+-----------+-------------+----------+--------+------+------------+---------+---------------+
| Table   | Non_unique | Key_name     | Seq_in_index | Column_name | Collation | Cardinality | Sub_part | Packed | Null | Index_type | Comment | Index_comment |
+---------+------------+--------------+--------------+-------------+-----------+-------------+----------+--------+------+------------+---------+---------------+
| student |          0 | PRIMARY      |            1 | id          | A         |           0 |     NULL |   NULL |      | BTREE      |         |               |
| student |          1 | name         |            1 | name        | A         |           0 |     NULL |   NULL | YES  | BTREE      |         |               |
| student |          1 | sex_index    |            1 | sex         | A         |           0 |     NULL |   NULL | YES  | BTREE      |         |               |
| student |          1 | id_name_index|            1 | id          | A         |           0 |     NULL |   NULL |      | BTREE      |         |               |
+---------+------------+--------------+--------------+-------------+-----------+-------------+----------+--------+------+------------+---------+---------------+
```

```
mysql> SHOW KEYS FROM student;
+---------+------------+--------------+--------------+-------------+-----------+-------------+----------+--------+------+------------+---------+---------------+
| Table   | Non_unique | Key_name     | Seq_in_index | Column_name | Collation | Cardinality | Sub_part | Packed | Null | Index_type | Comment | Index_comment |
+---------+------------+--------------+--------------+-------------+-----------+-------------+----------+--------+------+------------+---------+---------------+
| student |          0 | PRIMARY      |            1 | id          | A         |           0 |     NULL |   NULL |      | BTREE      |         |               |
| student |          1 | name         |            1 | name        | A         |           0 |     NULL |   NULL | YES  | BTREE      |         |               |
| student |          1 | sex_index    |            1 | sex         | A         |           0 |     NULL |   NULL | YES  | BTREE      |         |               |
| student |          1 | id_name_index|            1 | id          | A         |           0 |     NULL |   NULL |      | BTREE      |         |               |
+---------+------------+--------------+--------------+-------------+-----------+-------------+----------+--------+------+------------+---------+---------------+
```

① Table：代表表的名称。

② Non_unique：代表如果索引不能包括重复词，则为0，如果可以，则为1。

③ Key_name：代表索引的名称。

④ Seq_in_index：代表索引中的序列号，从1开始。

⑤ Column_name：代表列名称。

⑥ Collation：代表列以什么方式存储在索引中，A代表升序，NULL代表无分类。

⑦ Cardinality：代表索引中唯一值的数目的估计值，通过运行ANALYZE TABLE或myisamchk -a可以更新。基数根据被存储为整数的统计数据来计数，所以即使对于小型表，该值也没有必要是精确的。基数越大，当进行联合时，MySQL使用该索引的机会就越大。

⑧ Sub_part：代表如果列只是被部分地编入索引，则为被编入索引的字符的数目，如果整列被编入索引，则为NULL。

⑨ Packed：代表指示关键字如何被压缩，如果没有被压缩，则为NULL。

⑩ Null：代表如果列含有NULL，则含有YES，如果没有，则该列含有NO。

⑪ Index_type：代表索引的类型（BTREE, FULLTEXT, HASH, RTREE）。

⑫ Comment：代表注释。

7.1.3 删除索引

```
DROP INDEX index_name ON talbe_name
ALTER TABLE table_name DROP INDEX index_name
ALTER TABLE table_name DROP PRIMARY KEY
```

```
mysql> DROP INDEX sex_index ON student;
mysql> ALTER TABLE student DROP INDEX name;
mysql> ALTER TABLE student DROP PRIMARY KEY;
```

以上代码在删除PRIMARY KEY主键索引时使用，因为一个表只可能有一个PRIMARY KEY主键索引，所以不需要指定索引名。如果没有创建PRIMARY KEY索引，但表具有一个或多个UNIQUE索引，则将删除第一个UNIQUE索引。

如果从表中删除了某列，则索引会受到影响。对于多列组合的索引，如果删除其中的某列，则该列也会从索引中删除。如果删除组成索引的所有列，则整个索引将被删除。

下面给用户表user添加索引，给姓名name、性别sex添加索引，然后查看添加的索引，最后删除索引。

实战演练——给用户表user添加和删除索引

```
Microsoft Windows [版本 6.1.7601]
版权所有 (c) 2009 Microsoft Corporation。保留所有权利。

C:\Users\Administrator>mysql -uroot -p123456
mysql: [Warning] Using a password on the command line interface can be insecure.
Welcome to the MySQL monitor.  Commands end with ; or \g.
Your MySQL connection id is 3
Server version: 5.7.20 MySQL Community Server (GPL)

Copyright (c) 2000, 2017, Oracle and/or its affiliates. All rights reserved.

Oracle is a registered trademark of Oracle Corporation and/or its
affiliates. Other names may be trademarks of their respective
owners.

Type 'help;' or '\h' for help. Type '\c' to clear the current input statement.

mysql> USE shop;
Database changed

#查询用户表user
mysql> SELECT * FROM user;
+----+--------+------+------+----------+-------+-----------+--------+
| id | name   | sex  | age  | password | phone | loginName | remark |
+----+--------+------+------+----------+-------+-----------+--------+
|  3 | david  | 女   |   28 | 111111   | NULL  | david     | baann  |
|  4 | 小红   | 女   |   27 | 123456   | NULL  | xiaohong  | black  |
|  5 | 小明   | 男   |   10 | 123456   | NULL  | xiaoming  | berry  |
|  6 | 小刚   | 男   |   12 | 123456   | NULL  | xiaogang  | banner |
|  7 | 小王   | 男   |   14 | 111111   | NULL  | xiaowang  | banana |
|  8 | 小绿   | 女   |   34 | 222222   | NULL  | xiaolv    | car    |
|  9 | 晓峰   | 男   |   15 | 333333   | NULL  | xiaofeng  | carray |
| 10 | 小影   | 女   |   26 | 444444   | NULL  | xiaoying  | baaaa  |
| 11 | 大梅   | 女   |   27 | 555555   | NULL  | damei     | accc   |
| 15 | 小虎   | NULL | NULL | NULL     | NULL  | NULL      | NULL   |
+----+--------+------+------+----------+-------+-----------+--------+
10 rows in set (0.01 sec)

#给姓名name添加索引
```

```
mysql> ALTER TABLE user ADD INDEX nameIndex (name);
Query OK, 0 rows affected (0.43 sec)
Records: 0  Duplicates: 0  Warnings: 0
```

#给性别sex添加索引
```
mysql> ALTER TABLE user ADD INDEX sexIndex (sex);
Query OK, 0 rows affected (0.34 sec)
Records: 0  Duplicates: 0  Warnings: 0
```

#查询用户表user的索引
```
mysql> SHOW INDEX FROM user;
+-------+------------+-----------+--------------+-------------+-----------+-------------+----------+--------+------+------------+---------+---------------+
| Table | Non_unique | Key_name  | Seq_in_index | Column_name | Collation | Cardinality | Sub_part | Packed | Null | Index_type | Comment | Index_comment |
+-------+------------+-----------+--------------+-------------+-----------+-------------+----------+--------+------+------------+---------+---------------+
| user  |          0 | PRIMARY   |            1 | id          | A         |           9 |     NULL | NULL   |      | BTREE      |         |               |
| user  |          0 | id        |            1 | id          | A         |           9 |     NULL | NULL   |      | BTREE      |         |               |
| user  |          1 | idx_age   |            1 | age         | A         |           9 |     NULL | NULL   | YES  | BTREE      |         |               |
| user  |          1 | nameIndex |            1 | name        | A         |           9 |     NULL | NULL   | YES  | BTREE      |         |               |
| user  |          1 | sexIndex  |            1 | sex         | A         |           3 |     NULL | NULL   | YES  | BTREE      |         |               |
+-------+------------+-----------+--------------+-------------+-----------+-------------+----------+--------+------+------------+---------+---------------+
5 rows in set (0.05 sec)
```

#查询用户表user的索引
```
mysql> SHOW KEYS FROM user;
+-------+------------+-----------+--------------+-------------+-----------+-------------+----------+--------+------+------------+---------+---------------+
| Table | Non_unique | Key_name  | Seq_in_index | Column_name | Collation | Cardinality | Sub_part | Packed | Null | Index_type | Comment | Index_comment |
+-------+------------+-----------+--------------+-------------+-----------+-------------+----------+--------+------+------------+---------+---------------+
| user  |          0 | PRIMARY   |            1 | id          | A         |           9 |     NULL | NULL   |      | BTREE      |         |               |
| user  |          0 | id        |            1 | id          | A         |           9 |     NULL | NULL   |      | BTREE      |         |               |
| user  |          1 | idx_age   |            1 | age         | A         |           9 |     NULL | NULL   | YES  | BTREE      |         |               |
| user  |          1 | nameIndex |            1 | name        | A         |           9 |     NULL | NULL   | YES  | BTREE      |         |               |
| user  |          1 | sexIndex  |            1 | sex         | A         |           3 |     NULL | NULL   | YES  | BTREE      |         |               |
+-------+------------+-----------+--------------+-------------+-----------+-------------+----------+--------+------+------------+---------+---------------+
5 rows in set (0.00 sec)
```

#删除姓名name的索引
```
mysql> DROP INDEX nameIndex ON user;
Query OK, 0 rows affected (0.18 sec)
Records: 0  Duplicates: 0  Warnings: 0
```

#删除性别sex的索引
```
mysql> DROP INDEX sexIndex ON user;
Query OK, 0 rows affected (0.21 sec)
Records: 0  Duplicates: 0  Warnings: 0
```

#查看用户表user的索引
```
mysql> SHOW KEYS FROM user;
+-------+------------+----------+--------------+-------------+-----------+-------------+----------+--------+------+------------+---------+---------------+
| Table | Non_unique | Key_name | Seq_in_index | Column_name | Collation | Cardinality | Sub_part | Packed | Null | Index_type | Comment | Index_comment |
+-------+------------+----------+--------------+-------------+-----------+-------------+----------+--------+------+------------+---------+---------------+
```

```
+-------+------------+----------+--------------+------+-----+------+------+-------+-------+
| user  | 0          | PRIMARY  | 1            | id   | A   | 9    | NULL | NULL  |       | BTREE |   |   |
| user  | 0          | id       | 1            | id   | A   | 9    | NULL | NULL  |       | BTREE |   |   |
| user  | 1          | idx_age  | 1            | age  | A   | 9    | NULL | NULL  | YES   | BTREE |   |   |
+-------+------------+----------+--------------+------+-----+------+------+-------+-------+
3 rows in set (0.00 sec)

mysql>
```

7.2 常见的查询算法

查询算法有很多，有顺序查找算法、二分查找算法、二叉树查找算法、哈希查找算法，每个算法都有不同的应用场景，下面看看每种算法的含义和使用。

精讲视频

常见的查询算法

1. 顺序查找算法

顺序查找算法比较好理解，就是按顺序逐个查找。例如，有一组数据"2、3、5、7、8、9、10、16、21、25、30"，要从这组数据里查找"8"这个数据，使用顺序查找算法，就需要逐个比对数据，直到找到要查找的数据，对于大数据量，使用这种方式是低效的，它的时间复杂度为O(n)。

2. 二分查找算法

二分查找算法比顺序查找算法快，它的查找原理是从要查找的数据的中间元素开始。例如，有一组数据"2、3、5、7、8、9、10、16、21、25、30"，要从这组数据里查找"8"这个数据，使用二分查找算法，从中间元素"9"开始比对，以"9"为中心把数据分成两部分"2、3、5、7、8"和"10、16、21、25、30"，"8"是小于"9"的，所以在"2、3、5、7、8"这组数据里继续查找，同样使用二分查找算法再将其分成两部分"2、3"和"7、8"，"8"是大于"5"的，所以在"7、8"里查找数据，就是这样不断将数据分成两部分，在其中的一部分里查找。使用二分查找算法在一组数据里查找数据的前提是这组数据是有序的，它的时间复杂度为O(logn)。

3. 二叉树查找算法

二叉树具有如下特点：每个节点最多有两颗子树，节点的度最大为2；左子树和右子树是有顺序的，次序不能颠倒；即使某节点只有一个子树，也要区分左右子树；如图7.1所示。

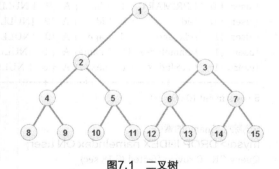

图7.1 二叉树

二叉树查找算法，先查找根节点，如果根节点的数值就是要查找的数值，直接返回根节点的数值；如果要查找的数值小于根节点的数值，则查找左子树，否则查找右子树。二叉树的高度是看树有几层，宽度是看哪一层节点最多，如图7.1所示的二叉树，高度是4，宽度是8，时间复杂度是O(log2N)。

4. 哈希查找算法

哈希查找算法是将数据值通过哈希函数创建一个哈希表，如果要查询某个数据值，也需要先通过哈希函数生成一个值，再到哈希表里去查询，如果查询到则返回数值。由于将通过哈希函数重新生成的数据值存放到哈希表中时很有可能重复，从而会产生冲突，冲突越少，查询得越快，如果没有冲突，则它的复杂度是O(1)。

7.3 索引的数据结构

MySQL索引是用来提高查询速度的。为了提高查询速度,需要有一个高效的算法和数据结构,而如果按顺序来查找某个数据,当数据量很大的时候,显然查询速度很慢。除了按顺序查找,也可以使用二分法来查找,但是二分法要求数据是有序的。另外,还可以使用二叉树方式来查找数据,但是它的叶子节点最多有两个,查询速度也会变慢。数据库常用数据结构B-Tree和B+Tree来构建数据库索引,MySQL数据库使用的就是B+Tree数据结构。

精讲视频

索引的数据结构

7.3.1 B-Tree数据结构

B-Tree数据结构称为平衡多路搜索树,它基于在二叉树的基础上采用多叉树,再使用平衡二叉树的思想。之所以采用多路搜索树,是因为这样可以降低树的深度,提高查询速度,如图7.2所示。

图7.2所示为一个深度为3的平衡多路二叉树,下面定义一条数据记录为一个二元组[key, data],key为记录的键值。对于不同的数据记录,key是互不相同的,data是用来描述B-Tree的数据结构的数据记录,但是它不包含key的数据。

(1)有一个根节点,根节点只有一个记录和两个孩子,如50就是根节点,下面有两个孩子。

(2)方框里的数字是key,线是指针,指针的末端是指向孩子节点。

(3)所有的叶子节点必须在同一层,也就是说它们具有相同的深度;每个叶子节点至

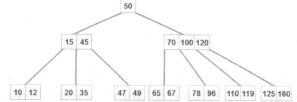

图7.2 平衡多路搜索树

少包含一个key和两个指针,最多包含2d-1个key和2d个指针,叶子节点的指针均为NULL。

(4)每个非叶子节点由n-1个key和n个指针组成,其中d≤n≤2d。

(5)在一个节点中,第n个子树中的所有key,小于这个节点中的第n个key,大于第n-1个key,如15、45这个子树,它的子树有3个范围:0~15、15~45、45以上。

B-Tree数据结构是先从根节点进行二分查找,如果小于根节点,就从左子树开始,从相应区间指针指向的节点递归进行查找;如果大于根节点,就从右子树递归进行查找,直到找到满足条件的节点,才返回对应节点的data。

7.3.2 B+Tree数据结构

B+Tree是B-Tree的变种,它也是一种平衡多路搜索树,MySQL就普遍使用B+Tree实现其索引结构。B+Tree数据结构如图7.3所示。

B+Tree和B-Tree比较起来有两点不同。

(1)非叶子节点指针数量与它的key的数量相同。

(2)非叶子节点只存储key,不存储data,叶子节点不存储指针。

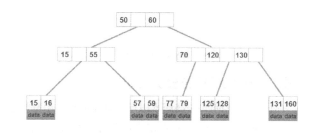

图7.3 B+Tree数据结构

7.4 索引实现原理

MySQL索引是存储引擎的内容，不同的存储引擎有不同的索引实现方式和原理，下面介绍常用的存储引擎的索引实现原理，包括MyISAM存储引擎、InnoDB存储引擎、MEMORY存储引擎。

7.4.1 MyISAM的索引实现

MyISAM存储引擎采用B+Tree数据结构作为索引的结构，它的叶子节点存放数据记录的地址，就如同书籍的目录，找到所要看的目录，按照目录指示的页码，找到对应的内容，也就是说MyISAM存储引擎的索引是先在B+Tree索引树上找到数据记录的地址，然后根据这个地址获取相应的数据。这样的搜索方式是高效的，在B+Tree索引树上很快就能找到地址，比一行一行地进行数据对比高效得多。MyISAM存储引擎的索引分为主索引和辅索引。

1. 主（唯一）索引

主索引是以主键或者唯一标识生成的索引，在创建表的时候，如果没有指定主（唯一）索引，存储引擎会自动维护一个主索引，它的每个节点是不允许重复的，就如同数据行的主键。

例如，有一个学生表，它有主键id、姓名name、性别sex、年龄age，那么它的主索引如图7.4所示。MyISAM主索引就是以主键为核心、采用B+Tree数据结构方式来构建索引树，叶子节点存放数据存储的地址，通过该地址可以快速找到该行数据。

2. 辅索引

辅索引就是以其他列构建的索引，而非以主键列构建的索引。它与主索引的区别就在于它允许重复，而主索引是不允许重复的；与主索引一样，它们都是在索引树的叶子节点存放地址，通过该地址找到数据。由于辅索引是允许重复的，通过地址可能找到多行数据，然后再通过条件找到满足的记录，如查找年龄为30岁，可以查找出多行记录，然后再查找名字是小刚，有可能就只有一条记录，也有可能有多条记录。

下面以年龄构建辅索引，如图7.5所示。

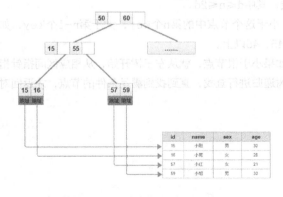

图7.4　MyISAM主索引

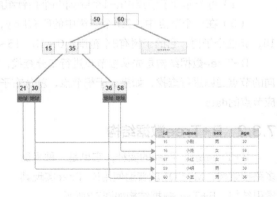

图7.5　MyISAM辅索引

7.4.2 InnoDB的索引实现

MySQL数据库在5.5版本以后，默认的存储引擎就是InnoDB存储引擎。它也是采用B+Tree数据结构来构建索引的，但是和MyISAM存储引擎的索引数据结构不同，MyISAM存储引擎将数据记录的地址放置在索引树的叶子节点下，通过该地址就可以找到存储的数据，而InnoDB存储引擎的索引是叶子节点直接存放数据，找到该叶子节点也就是找到数据，它分为主索引和二级索引。

1. 主（唯一）索引

主索引是主键或者唯一列来构建的索引，如果没有指定主键或者唯一列，MySQL自己维护一套主索引，它是不允许重复的。

例如，有一个学生表，它有主键id、姓名name、性别sex、年龄age，那么它的主索引如图7.6所示。

2. 二级索引

InnoDB的二级索引是以其他列而非主键列构建的索引，所以它是允许重复的。它的叶子节点存放主索引的值，通过主索引再直接找到数据，这样就不用在每棵索引树上存储数据，这样就节省了存储空间，同时便于维护。

下面以年龄age构建二级索引，它的叶子节点存放主索引的值，如图7.7所示。

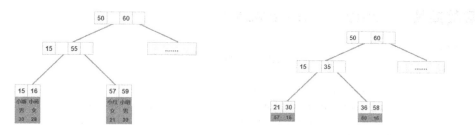

图7.6　InnoDB主索引　　　　　　图7.7　InnoDB二级索引

7.4.3　MEMORY的索引实现

MEMORY引擎又称为HEAP引擎，用于创建特殊用途的表且内容存储在内存中。将数据存储在内存中，能够实现快速访问和低延迟。它的默认索引方式采用Hash索引的方式来构建。Hash索引是通过Hash函数计算出来的结果，然后将该结果随机地存放到硬盘上。

例如，有一个学生表，它有主键id、姓名name、性别sex、年龄age等列，以主键id来构建Hash索引，通过Hash函数hash(id)来计算出一个结果值，将该结果值随机地存放到硬盘中，如图7.8所示。

从7.8图中可以看出，通过hsah()函数，对某列要加索引重新计算出一个结果值随机记录到磁盘上，如果计算出的结果值没有重复的，那么它的时间复杂度就是O(1)。随着数据量的增大，通过hash函数计算出的结果值很可能重复，这样查询效率会逐渐下降，但是它的检索速度也是非常快的。

图7.8　Hash索引

虽然Hash索引的检索速度非常快，但是它也有使用的局限性，它不会对范围查询进行优化，如查找一个时间段内的数据，它就做不到；它也无法利用前缀索引，这个前缀索引我们在后面的章节中会进行介绍；它也无法对排序进行优化；它查询时必须通过Hash函数来计算结果找到数据存放的位置，然后再到表里进行查询。

7.5　索引的应用

下面通过学生表student来学习索引的应用，包括创建表、添加索引、分析索引的使用等内容。

精讲视频

索引的应用

7.5.1　创建表及添加索引

创建一个学生表student，它有主键id列、姓名name列、性别sex列、年龄age列、年级grade列、班级class列、学号num列，如表7.1所示，然后给id、name、sex、age每列单独添加索引。

表7.1 student表

id	name	sex	age	grade	class	num
1	刘明	男	19	高三	6班	3001
2	吴倩	女	18	高三	6班	3002
3	张欣	男	17	高二	1班	2001
4	孙晓	女	17	高二	1班	2002
5	赵英俊	男	16	高一	2班	1001
6	柳师师	女	17	高一	2班	1002

实战演练——创建学生表student并添加索引

```
Microsoft Windows [版本 6.1.7601]
版权所有 (c) 2009 Microsoft Corporation。保留所有权利。

C:\Users\Administrator>mysql -uroot -p123456
mysql: [Warning] Using a password on the command line interface can be insecure.
Welcome to the MySQL monitor.  Commands end with ; or \g.
Your MySQL connection id is 3
Server version: 5.7.20 MySQL Community Server (GPL)

Copyright (c) 2000, 2017, Oracle and/or its affiliates. All rights reserved.

Oracle is a registered trademark of Oracle Corporation and/or its
affiliates. Other names may be trademarks of their respective
owners.

Type 'help;' or '\h' for help. Type '\c' to clear the current input statement.

mysql> USE shop;
Database changed

#删除student表
mysql> drop table student;
Query OK, 0 rows affected (0.15 sec)

#创建student表
mysql> CREATE TABLE student(
    -> id int not null AUTO_INCREMENT,
    -> name varchar(255),
    -> sex varchar(5),
    -> age varchar(5),
    -> grade varchar(10),
    -> class varchar(10),
    -> num int,
    -> primary key(id)
    -> );
Query OK, 0 rows affected (0.42 sec)

#插入数据
```

```
mysql> insert into student values(1,'刘明','男','19','高三','6班',3001);
Query OK, 1 row affected (0.09 sec)

#插入数据
mysql> insert into student values(2,'吴倩','女','18','高三','6班',3002);
Query OK, 1 row affected (0.00 sec)

#插入数据
mysql> insert into student values(3,'张欣','男','17','高二','1班',2001);
Query OK, 1 row affected (0.00 sec)

#插入数据
mysql> insert into student values(4,'孙晓','女','17','高二','1班',2002);
Query OK, 1 row affected (0.00 sec)

#插入数据
mysql> insert into student values(5,'赵英俊','男','16','高一','2班',1001);
Query OK, 1 row affected (0.00 sec)

#插入数据
mysql> insert into student values(6,'柳师师','女','17','高一','2班',1002);
Query OK, 1 row affected (0.00 sec)

#查询数据
mysql> SELECT * FROM student;
+----+--------+-----+-----+-------+-------+------+
| id | name   | sex | age | grade | class | num  |
+----+--------+-----+-----+-------+-------+------+
|  1 | 刘明   | 男  | 19  | 高三  | 6班   | 3001 |
|  2 | 吴倩   | 女  | 18  | 高三  | 6班   | 3002 |
|  3 | 张欣   | 男  | 17  | 高二  | 1班   | 2001 |
|  4 | 孙晓   | 女  | 17  | 高二  | 1班   | 2002 |
|  5 | 赵英俊 | 男  | 16  | 高一  | 2班   | 1001 |
|  6 | 柳师师 | 女  | 17  | 高一  | 2班   | 1002 |
+----+--------+-----+-----+-------+-------+------+
6 rows in set (0.00 sec)

#添加唯一索引
mysql> ALTER TABLE student ADD UNIQUE (id);
Query OK, 0 rows affected (0.70 sec)
Records: 0  Duplicates: 0  Warnings: 0

#添加索引
mysql> ALTER TABLE student ADD INDEX name_index (name);
Query OK, 0 rows affected (0.30 sec)
Records: 0  Duplicates: 0  Warnings: 0

#添加索引
mysql> ALTER TABLE student ADD INDEX sex_index (sex);
Query OK, 0 rows affected (0.63 sec)
Records: 0  Duplicates: 0  Warnings: 0
```

```
#添加索引
mysql> ALTER TABLE student ADD INDEX age_index (age);
Query OK, 0 rows affected (0.42 sec)
Records: 0  Duplicates: 0  Warnings: 0

#查看索引
mysql> SHOW INDEX FROM student;
+-------+------+------------+----------+-------------+-----------+-------------+----------+--------+------+------------+---------+---------------+
| Table | Non_unique | Key_name | Seq_in_index | Column_name | Collation | Cardinality | Sub_part | Packed | Null | Index_type | Comment | Index_comment |
+-------+------+------------+----------+-------------+-----------+-------------+----------+--------+------+------------+---------+---------------+
| student | 0 | PRIMARY | 1 | id | A | 0 | NULL | NULL |  | BTREE |  |  |
| student | 0 | id | 1 | id | A | 0 | NULL | NULL | YES | BTREE |  |  |
| student | 1 | name_index | 1 | name | A |  | NULL | NULL | YES | BTREE |  |  |
| student | 1 | sex_index | 1 | sex | A |  | NULL | NULL | YES | BTREE |  |  |
| student | 1 | age_index | 1 | age | A |  | NULL | NULL | YES | BTREE |  |  |
+-------+------+------------+----------+-------------+-----------+-------------+----------+--------+------+------------+---------+---------------+
mysql>
```

7.5.2　explain分析索引

在student表的id、name、sex、age 4列上分别加上索引，怎么知道使用哪个索引呢？我们使用explain来分析索引。

下面从student表里查询姓名为刘明、性别为男的学生。

SELECT * FROM student WHERE name='刘明' and sex='男';

explain SELECT * FROM student WHERE name='刘明' and sex='男' \G

实战演练——用explain分析student表索引

```
Microsoft Windows [版本 6.1.7601]
版权所有 (c) 2009 Microsoft Corporation。保留所有权利。

C:\Users\Administrator>mysql –uroot –p123456
mysql: [Warning] Using a password on the command line interface can be insecure.
Welcome to the MySQL monitor.  Commands end with ; or \g.
Your MySQL connection id is 6
Server version: 5.7.20 MySQL Community Server (GPL)

Copyright (c) 2000, 2017, Oracle and/or its affiliates. All rights reserved.

Oracle is a registered trademark of Oracle Corporation and/or its
affiliates. Other names may be trademarks of their respective
owners.

Type 'help;' or '\h' for help. Type '\c' to clear the current input statement.

mysql> USE shop;
Database changed
mysql> SELECT * FROM student;
```

```
+----+--------+------+------+--------+--------+------+
| id | name   | sex  | age  | grade  | class  | num  |
+----+--------+------+------+--------+--------+------+
|  1 | 刘明   | 男   | 19   | 高三   | 6班    | 3001 |
|  2 | 吴倩   | 女   | 18   | 高三   | 6班    | 3002 |
|  3 | 张欣   | 男   | 17   | 高二   | 1班    | 2001 |
|  4 | 孙晓   | 女   | 17   | 高二   | 1班    | 2002 |
|  5 | 赵英俊 | 男   | 16   | 高一   | 2班    | 1001 |
|  6 | 柳师师 | 女   | 17   | 高一   | 2班    | 1002 |
+----+--------+------+------+--------+--------+------+
6 rows in set (0.00 sec)

mysql> SELECT * FROM student WHERE name='刘明' and sex='男';
+----+------+------+------+--------+--------+------+
| id | name | sex  | age  | grade  | class  | num  |
+----+------+------+------+--------+--------+------+
|  1 | 刘明 | 男   | 19   | 高三   | 6班    | 3001 |
+----+------+------+------+--------+--------+------+
1 row in set (0.09 sec)

mysql> explain SELECT * FROM student WHERE name='刘明' and sex='男' \G
*************************** 1. row ***************************
           id: 1
  select_type: SIMPLE
        table: student
   partitions: NULL
         type: ref
possible_keys: name_index,sex_index
          key: name_index
      key_len: 768
          ref: const
         rows: 1
     filtered: 50.00
        Extra: Using where
1 row in set, 1 warning (0.00 sec)

mysql>
```

explain用来分析SQL语句的查询情况,它包含id、select_type、table、type、possible_keys、key、key_len、ref、rows、Extra字段,字段说明如表7.2所示。

表7.2 explain分析SQL语句的字段说明

序号	字段名称	字 段 含 义
1	id	用来标识SELECT语句的编号,有多个select就会有多个id,编号从1开始递增
2	select_type	SIMPLE:简单查询,不包含子查询; PRIMARY:主查询,也就是最外层的SELECT查询; SUBQUERY:子查询中的第一个SELECT查询,不依赖于外部查询的结果集; UNION:UNION 中的第二个或随后的SELECT查询,不依赖于外部查询的结果集; UNION RESULT:UNION的结果集; DEPENDENT UNION:UNION语句中的第二个SELECT,依赖于外部子查询; DEPENDENT SUBQUERY:子查询中的第一个SELECT,依赖于外面的查询

续表

序号	字段名称	字 段 含 义
3	table	查询所使用到的表名，它可以是实际的表名，也可以是表的别名
4	type	显示连接使用的类型，可以决定查询速度，下面按最优到最差的类型排序。 system：表仅有一行（=系统表）； const：用于常数值比较PRIMARY KEY时，当查询的表仅有一行时，使用system； eq_ref：相等引用查询，并且只能找到一行数据； ref：引用查询，快速查找到多个符合条件的行； ref_or_null：如同ref，必须在初次查找的结果里找出null条目，然后进行二次查找； index_merge：索引合并优化被使用； unique_subquery：在某些 IN 查询中使用此种类型，查询的是唯一索引； index_subquery：与unique_subquery用法类似，只是查询的是非唯一索引； range：检索给定范围的行； index：全表扫描索引树的节点； all：从头到尾全表扫描，它是扫描全表数据
5	possible_keys	可能用到的索引，最终只能使用一个索引，如果为空，说明没有可用的索引
6	key	实际用到的索引，只能使用一个索引，如果为 NULL，则没有使用索引
7	key_len	用到的索引长度，在不损失精确性的情况下，长度越短越好
8	ref	连接查询时，表与表之间的连接关系
9	rows	扫描的行数
10	Extra	附加信息，查看索引的使用情况。 Using index：使用索引，没有查询数据表，只用索引表完成查询，也叫覆盖索引，如果同时出现Using where，代表使用索引来查找记录，但是需要查询到数据表； Using where：条件查询，如果不读取表的所有数据或者通过索引就可以获取所有需要的数据，则会出现 Using where； Using filesort：文件排序是一种排序策略，使用ORDER BY排序语句会出现该信息； Using temporary：为了得到结果，使用了临时表

1. id查询编号

id是用来标识SELECT语句的编号，有多个SELECT语句就会有多个id，但是如果查询的SELECT语句是相同的select_type查询类型和相同的table表，其id也不会自增，因为它认为是一条语句。

```
#id自增语句
mysql> explain SELECT * FROM student WHERE id = (SELECT id FROM student WHERE id = 3) \G
*************************** 1. row ***************************
           id: 1
  select_type: PRIMARY
        table: student
   partitions: NULL
         type: const
possible_keys: PRIMARY,id
          key: PRIMARY
      key_len: 4
          ref: const
         rows: 1
     filtered: 100.00
        Extra: NULL
*************************** 2. row ***************************
           id: 2
```

```
        select_type: SUBQUERY
              table: student
         partitions: NULL
               type: const
      possible_keys: PRIMARY,id
                key: PRIMARY
            key_len: 4
                ref: const
               rows: 1
           filtered: 100.00
              Extra: Using index
2 rows in set, 1 warning (0.08 sec)
```

#id不自增语句
```
mysql> explain SELECT * FROM student WHERE id in (SELECT id FROM student WHERE id = 3) \G
*************************** 1. row ***************************
                 id: 1
        select_type: SIMPLE
              table: student
         partitions: NULL
               type: const
      possible_keys: PRIMARY,id
                key: PRIMARY
            key_len: 4
                ref: const
               rows: 1
           filtered: 100.00
              Extra: NULL
*************************** 2. row ***************************
                 id: 1
        select_type: SIMPLE
              table: student
         partitions: NULL
               type: const
      possible_keys: PRIMARY,id
                key: PRIMARY
            key_len: 4
                ref: const
               rows: 1
           filtered: 100.00
              Extra: Using index
2 rows in set, 1 warning (0.00 sec)
```

2. select_type查询类型

select_type是查询类型，它分为以下几种：SIMPLE为简单查询，不包含子查询；PRIMARY为主查询，也就是最外层的SELECT查询；SUBQUERY为子查询中的第一个SELECT查询，不依赖于外部查询的结果集；UNION为UNION语句中的第二个或随后的SELECT查询，不依赖于外部查询的结果集；UNION RESULT为UNION的结果集；DEPENDENT UNION为UNION语句中的第二个SELECT，依赖于外部子查询。

（1）简单查询SIMPLE。

```
mysql> explain SELECT * FROM student \G
*************************** 1. row ***************************
```

```
                 id: 1
       select_type: SIMPLE
            table: student
       partitions: NULL
             type: ALL
    possible_keys: NULL
              key: NULL
          key_len: NULL
              ref: NULL
             rows: 6
         filtered: 100.00
            Extra: NULL
1 row in set, 1 warning (0.00 sec)
```

（2）主查询PRIMARY、子查询SUBQUERY。

```
mysql> explain SELECT * FROM student WHERE id = (SELECT id FROM student WHERE id = 3) \G
*************************** 1. row ***************************
                 id: 1
       select_type: PRIMARY
            table: student
       partitions: NULL
             type: const
    possible_keys: PRIMARY,id
              key: PRIMARY
          key_len: 4
              ref: const
             rows: 1
         filtered: 100.00
            Extra: NULL
*************************** 2. row ***************************
                 id: 2
       select_type: SUBQUERY
            table: student
       partitions: NULL
             type: const
    possible_keys: PRIMARY,id
              key: PRIMARY
          key_len: 4
              ref: const
             rows: 1
         filtered: 100.00
            Extra: Using index
2 rows in set, 1 warning (0.00 sec)
```

（3）UNION查询，在UNION之后的SELECT查询，不依赖于外部子查询。

```
mysql> explain SELECT * FROM student WHERE id=2 UNION SELECT * FROM student WHERE id=3 \G
*************************** 1. row ***************************
                 id: 1
       select_type: PRIMARY
            table: student
       partitions: NULL
             type: const
    possible_keys: PRIMARY,id
```

```
            key: PRIMARY
        key_len: 4
            ref: const
           rows: 1
       filtered: 100.00
          Extra: NULL
*************************** 2. row ***************************
             id: 2
    select_type: UNION
          table: student
     partitions: NULL
           type: const
  possible_keys: PRIMARY,id
            key: PRIMARY
        key_len: 4
            ref: const
           rows: 1
       filtered: 100.00
          Extra: NULL
*************************** 3. row ***************************
             id: NULL
    select_type: UNION RESULT
          table: <union1,2>
     partitions: NULL
           type: ALL
  possible_keys: NULL
            key: NULL
        key_len: NULL
            ref: NULL
           rows: NULL
       filtered: NULL
          Extra: Using temporary
3 rows in set, 1 warning (0.00 sec)

mysql>
```

（4）DEPENDENT UNION查询，DEPENDENT SUBQUERY查询。

```
mysql> explain SELECT * FROM student WHERE id IN (SELECT id FROM student WHERE id=2 UNION SELECT id FROM student WHERE id=3) \G
*************************** 1. row ***************************
             id: 1
    select_type: PRIMARY
          table: student
     partitions: NULL
           type: ALL
  possible_keys: NULL
            key: NULL
        key_len: NULL
            ref: NULL
           rows: 6
       filtered: 100.00
          Extra: Using where
```

```
*************************** 2. row ***************************
           id: 2
  select_type: DEPENDENT SUBQUERY
        table: student
   partitions: NULL
         type: const
possible_keys: PRIMARY,id
          key: PRIMARY
      key_len: 4
          ref: const
         rows: 1
     filtered: 100.00
        Extra: Using index
*************************** 3. row ***************************
           id: 3
  select_type: DEPENDENT UNION
        table: student
   partitions: NULL
         type: const
possible_keys: PRIMARY,id
          key: PRIMARY
      key_len: 4
          ref: const
         rows: 1
     filtered: 100.00
        Extra: Using index
*************************** 4. row ***************************
           id: NULL
  select_type: UNION RESULT
        table: <union2,3>
   partitions: NULL
         type: ALL
possible_keys: NULL
          key: NULL
      key_len: NULL
          ref: NULL
         rows: NULL
     filtered: NULL
        Extra: Using temporary
4 rows in set, 1 warning (0.07 sec)

mysql>
```

3. table表名

table是查询的表名，它可以是实际的表名，可以是表的别名，也可以为NULL。

```
#实际表名
mysql> explain SELECT * FROM student \G
*************************** 1. row ***************************
           id: 1
  select_type: SIMPLE
        table: student
   partitions: NULL
```

```
        type: ALL
possible_keys: NULL
          key: NULL
      key_len: NULL
          ref: NULL
         rows: 6
     filtered: 100.00
        Extra: NULL
1 row in set, 1 warning (0.00 sec)

#表的别名
mysql> explain SELECT * FROM student as temp \G
*************************** 1. row ***************************
           id: 1
  select_type: SIMPLE
        table: temp
   partitions: NULL
         type: ALL
possible_keys: NULL
          key: NULL
      key_len: NULL
          ref: NULL
         rows: 6
     filtered: 100.00
        Extra: NULL
1 row in set, 1 warning (0.00 sec)

#表名为NULL
mysql> explain SELECT 1 \G
*************************** 1. row ***************************
           id: 1
  select_type: SIMPLE
        table: NULL
   partitions: NULL
         type: NULL
possible_keys: NULL
          key: NULL
      key_len: NULL
          ref: NULL
         rows: NULL
     filtered: NULL
        Extra: No tables used
1 row in set, 1 warning (0.00 sec)

mysql>
```

4. type扫描类型

Type是用在查询过程中的扫描方式，不同的扫描方式可以决定查询的速度。下面按最优到最差的顺序介绍扫描类型。

system：表仅有一行（=系统表）。

const：用于常数值比较主键或者唯一索引时，当查询的表仅有一行时，使用system。

eq_ref：相等引用查询，并且只能找到一行数据，每次与之前的表合并都只在该表读取一行，这是除了 system、const 之外最好的一种，它使用=，而且索引的所有部分都参与join且索引是主键或非空唯一键的索引。

ref：引用查询，能快速查找到多个符合条件的行，如果每次只匹配少数行，那就是比较好的一种，使用=或<=>，可以是左覆盖索引、非主键或非唯一键索引。

ref_or_null：与ref类似，必须在初次查找的结果里找出null条目，然后进行二次查找。

index_merge：索引合并优化被使用，包括交集、并集以及交集之间的并集，但不包括跨表和全文索引。

unique_subquery：在某些 IN 查询中使用此种类型，查询的是唯一索引。

index_subquery：与unique_subquery类似，查询的是非唯一索引。

range：检索给定范围的行。

index：全表扫描索引树的节点。

all：从头到尾全表扫描，它是扫描全表数据。

（1）type=all，扫描全表数据，速度最慢。

```
mysql> explain SELECT * FROM student \G
*************************** 1. row ***************************
           id: 1
  select_type: SIMPLE
        table: student
   partitions: NULL
         type: ALL
possible_keys: NULL
          key: NULL
      key_len: NULL
          ref: NULL
         rows: 6
     filtered: 100.00
        Extra: NULL
1 row in set, 1 warning (0.00 sec)

mysql>
```

（2）type=index，扫描索引所有节点。

```
mysql> explain SELECT name FROM student \G
*************************** 1. row ***************************
           id: 1
  select_type: SIMPLE
        table: student
   partitions: NULL
         type: index
possible_keys: NULL
          key: name_index
      key_len: 768
          ref: NULL
         rows: 6
     filtered: 100.00
        Extra: Using index
1 row in set, 1 warning (0.00 sec)

mysql>
```

（3）type=range，扫描给定范围的行。

```
mysql> explain SELECT name FROM student WHERE id > 2 \G
*************************** 1. row ***************************
           id: 1
  select_type: SIMPLE
        table: student
   partitions: NULL
         type: range
possible_keys: PRIMARY,id
          key: PRIMARY
      key_len: 4
          ref: NULL
         rows: 4
     filtered: 100.00
        Extra: Using where
1 row in set, 1 warning (0.06 sec)

mysql>
```

（4）type=ref，非唯一索引扫描，返回匹配某个单独值的所有行。

```
mysql> explain SELECT * FROM student WHERE name='小明' \G
*************************** 1. row ***************************
           id: 1
  select_type: SIMPLE
        table: student
   partitions: NULL
         type: ref
possible_keys: name_index
          key: name_index
      key_len: 768
          ref: const
         rows: 1
     filtered: 100.00
        Extra: NULL
1 row in set, 1 warning (0.00 sec)
```

（5）type=eq_ref，唯一索引扫描，只能返回满足条件的一行数据。

```
mysql> explain SELECT * FROM student WHERE id IN (SELECT id FROM student) \G
*************************** 1. row ***************************
           id: 1
  select_type: SIMPLE
        table: student
   partitions: NULL
         type: ALL
possible_keys: PRIMARY,id
          key: NULL
      key_len: NULL
          ref: NULL
         rows: 6
     filtered: 100.00
        Extra: NULL
*************************** 2. row ***************************
           id: 1
  select_type: SIMPLE
```

```
            table: student
       partitions: NULL
             type: eq_ref
    possible_keys: PRIMARY,id
              key: PRIMARY
          key_len: 4
              ref: shop.student.id
             rows: 1
         filtered: 100.00
            Extra: Using index
2 rows in set, 1 warning (0.00 sec)

mysql>
```

（6）type=const、system，以常量的方式进行扫描。

```
mysql> explain SELECT * FROM student WHERE id=1 \G
*************************** 1. row ***************************
               id: 1
      select_type: SIMPLE
            table: student
       partitions: NULL
             type: const
    possible_keys: PRIMARY,id
              key: PRIMARY
          key_len: 4
              ref: const
             rows: 1
         filtered: 100.00
            Extra: NULL
1 row in set, 1 warning (0.00 sec)
```

（7）type=NULL，不用访问表或者索引。

```
mysql> explain SELECT NOW() \G
*************************** 1. row ***************************
               id: 1
      select_type: SIMPLE
            table: NULL
       partitions: NULL
             type: NULL
    possible_keys: NULL
              key: NULL
          key_len: NULL
              ref: NULL
             rows: NULL
         filtered: NULL
            Extra: No tables used
1 row in set, 1 warning (0.05 sec)

mysql>
```

5. possible_keys、key、key_len

possible_keys把可能用到的索引都罗列出来，最终只能使用一个索引，如果为空，说明没有可用的索引；key指明实际用到的索引，只能使用一个索引，如果为NULL，则没有使用索引；key_len指明用到的索引长度，在不损失精确性的情况下，长度越短越好。

```
#没有用到索引
mysql> explain SELECT * FROM student \G
*************************** 1. row ***************************
           id: 1
  select_type: SIMPLE
        table: student
   partitions: NULL
         type: ALL
possible_keys: NULL
          key: NULL
      key_len: NULL
          ref: NULL
         rows: 6
     filtered: 100.00
        Extra: NULL
1 row in set, 1 warning (0.00 sec)

#用到索引
mysql> explain SELECT * FROM student WHERE id = 3 \G
*************************** 1. row ***************************
           id: 1
  select_type: SIMPLE
        table: student
   partitions: NULL
         type: const
possible_keys: PRIMARY,id
          key: PRIMARY
      key_len: 4
          ref: const
         rows: 1
     filtered: 100.00
        Extra: NULL
1 row in set, 1 warning (0.00 sec)

mysql>
```

6. ref连接关系

ref用来说明表与表连接查询时是如何建立连接关系的。

```
mysql> explain SELECT * FROM student s1 INNER JOIN student s2 on s1.id=s2.id \G
*************************** 1. row ***************************
           id: 1
  select_type: SIMPLE
        table: s1
   partitions: NULL
         type: ALL
possible_keys: PRIMARY,id
          key: NULL
      key_len: NULL
          ref: NULL
         rows: 6
     filtered: 100.00
        Extra: NULL
*************************** 2. row ***************************
```

```
            id: 1
  select_type: SIMPLE
        table: s2
   partitions: NULL
         type: eq_ref
possible_keys: PRIMARY,id
          key: PRIMARY
      key_len: 4
          ref: shop.s1.id
         rows: 1
     filtered: 100.00
        Extra: NULL
2 rows in set, 1 warning (0.00 sec)

mysql>
```

7. rows扫描的行数

rows用来估计要扫描的行数。

```
mysql> explain SELECT * FROM student \G
*************************** 1. row ***************************
            id: 1
  select_type: SIMPLE
        table: student
   partitions: NULL
         type: ALL
possible_keys: NULL
          key: NULL
      key_len: NULL
          ref: NULL
         rows: 6
     filtered: 100.00
        Extra: NULL
1 row in set, 1 warning (0.00 sec)

mysql> explain SELECT * FROM student WHERE id=4 \G
*************************** 1. row ***************************
            id: 1
  select_type: SIMPLE
        table: student
   partitions: NULL
         type: const
possible_keys: PRIMARY,id
          key: PRIMARY
      key_len: 4
          ref: const
         rows: 1
     filtered: 100.00
        Extra: NULL
1 row in set, 1 warning (0.00 sec)
```

8. Extra附加信息

Extra附加信息有4个选项值：Using index、Using where、Using filesort、Using temporary。Using index表示使用了索引就可以查询到想要的内容，也叫覆盖索引，查询效率最高；如果仅使用索引还不能完成查询,就会出现Using where,使用索引来查找记录,同时也需要到数据表中进行查询;Using filesort是一种排序策略,通常在使用到排序语句ORDER BY的时候,会出现该信息;Using temporary使用了临时表,查询效率较低。

```
#没有附加信息
mysql> explain SELECT grade FROM student \G
*************************** 1. row ***************************
           id: 1
  select_type: SIMPLE
        table: student
   partitions: NULL
         type: ALL
possible_keys: NULL
          key: NULL
      key_len: NULL
          ref: NULL
         rows: 6
     filtered: 100.00
        Extra: NULL
1 row in set, 1 warning (0.00 sec)

#使用Using index,使用索引查询
mysql> explain SELECT id FROM student WHERE id=4 \G
*************************** 1. row ***************************
           id: 1
  select_type: SIMPLE
        table: student
   partitions: NULL
         type: const
possible_keys: PRIMARY,id
          key: PRIMARY
      key_len: 4
          ref: const
         rows: 1
     filtered: 100.00
        Extra: Using index
1 row in set, 1 warning (0.00 sec)

#使用Using index,使用索引查询,同时使用Using where进行条件判断
mysql> explain SELECT id FROM student WHERE id>4 \G
*************************** 1. row ***************************
           id: 1
  select_type: SIMPLE
        table: student
   partitions: NULL
         type: range
possible_keys: PRIMARY,id
          key: id
      key_len: 4
          ref: NULL
         rows: 2
     filtered: 100.00
```

```
        Extra: Using where; Using index
1 row in set, 1 warning (0.00 sec)
```

#使用Using where进行条件判断
```
mysql> explain SELECT * FROM student WHERE id>4 \G
*************************** 1. row ***************************
           id: 1
  select_type: SIMPLE
        table: student
   partitions: NULL
         type: range
possible_keys: PRIMARY,id
          key: PRIMARY
      key_len: 4
          ref: NULL
         rows: 2
     filtered: 100.00
        Extra: Using where
1 row in set, 1 warning (0.00 sec)
```

#使用Using where进行条件判断，同时进行文件排序Using filesort
```
mysql> explain SELECT * FROM student WHERE sex='男' ORDER BY age DESC \G
*************************** 1. row ***************************
           id: 1
  select_type: SIMPLE
        table: student
   partitions: NULL
         type: ALL
possible_keys: sex_index
          key: NULL
      key_len: NULL
          ref: NULL
         rows: 6
     filtered: 50.00
        Extra: Using where; Using filesort
1 row in set, 1 warning (0.00 sec)
```

#使用临时表Using temporary，同时进行文件排序Using filesort
```
mysql> explain SELECT * FROM student UNION SELECT * FROM student ORDER BY id DESC\G
*************************** 1. row ***************************
           id: 1
  select_type: PRIMARY
        table: student
   partitions: NULL
         type: ALL
possible_keys: NULL
          key: NULL
      key_len: NULL
          ref: NULL
         rows: 6
     filtered: 100.00
        Extra: NULL
```

```
*************************** 2. row ***************************
           id: 2
  select_type: UNION
        table: student
   partitions: NULL
         type: ALL
possible_keys: NULL
          key: NULL
      key_len: NULL
          ref: NULL
         rows: 6
     filtered: 100.00
        Extra: NULL
*************************** 3. row ***************************
           id: NULL
  select_type: UNION RESULT
        table: <union1,2>
   partitions: NULL
         type: ALL
possible_keys: NULL
          key: NULL
      key_len: NULL
          ref: NULL
         rows: NULL
     filtered: NULL
        Extra: Using temporary; Using filesort
3 rows in set, 1 warning (0.00 sec)
```

7.5.3 索引使用策略

索引的使用，可以分为匹配全值索引查询、匹配最左前缀索引查询、匹配列前缀索引查询、匹配值的范围索引查询、仅对索引进行查询（索引覆盖）。

（1）匹配全值索引查询：针对建索引的列可以匹配列的全值进行查询。

```
mysql> explain SELECT * FROM student WHERE id = 3 \G
*************************** 1. row ***************************
           id: 1
  select_type: SIMPLE
        table: student
   partitions: NULL
         type: const
possible_keys: PRIMARY,id
          key: PRIMARY
      key_len: 4
          ref: const
         rows: 1
     filtered: 100.00
        Extra: NULL
1 row in set, 1 warning (0.00 sec)

mysql>
```

（2）匹配最左前缀索引查询：索引不仅可以进行单列索引，也可以建立联合索引，也就是把多列建成一个联合索引。

例如，把学生表中的姓名name、性别sex、年龄age建立成联合索引，按照最左前缀匹配原则，它可以匹配name单独一列，可以匹配name、sex两列，可以匹配name、sex、age三列，但是不能匹配sex或者age单独列，不能匹配name、age列或者sex、age列。

```
#把姓名name、性别sex、年龄age建立成联合索引
mysql> ALTER TABLE student ADD INDEX nsa_index (name,sex,age);
Query OK, 0 rows affected (0.35 sec)
Records: 0  Duplicates: 0  Warnings: 0

#删除姓名name列索引
mysql>  DROP INDEX name_index ON student;
Query OK, 0 rows affected (0.20 sec)
Records: 0  Duplicates: 0  Warnings: 0

#按姓名name索引查找，匹配最左前缀列
mysql> explain SELECT * FROM student WHERE name = '小明' \G
*************************** 1. row ***************************
           id: 1
  select_type: SIMPLE
        table: student
   partitions: NULL
         type: ref
possible_keys: nsa_index
          key: nsa_index
      key_len: 768
          ref: const
         rows: 1
     filtered: 100.00
        Extra: NULL
1 row in set, 1 warning (0.00 sec)

#当可选的索引有多个时，sex_index，nsa_index，存储引擎会选择最优的索引
mysql> explain SELECT * FROM student WHERE name = '小明' AND sex='男'\G
*************************** 1. row ***************************
           id: 1
  select_type: SIMPLE
        table: student
   partitions: NULL
         type: ref
possible_keys: sex_index,nsa_index
          key: nsa_index
      key_len: 786
          ref: const,const
         rows: 1
     filtered: 100.00
        Extra: NULL
1 row in set, 1 warning (0.00 sec)

#当可选的索引有多个时，age_index，nsa_index，存储引擎会选择最优的索引
mysql> explain SELECT * FROM student WHERE name = '小明' AND age='30'\G
*************************** 1. row ***************************
           id: 1
```

```
       select_type: SIMPLE
             table: student
        partitions: NULL
              type: ref
     possible_keys: age_index,nsa_index
               key: age_index
           key_len: 18
               ref: const
              rows: 1
          filtered: 16.67
             Extra: Using where
1 row in set, 1 warning (0.00 sec)
```

#当可选的索引有多个时，存储引擎会选择最优的索引
```
mysql> explain SELECT * FROM student WHERE name='小明' AND sex='男' AND age='30'\G
*************************** 1. row ***************************
                id: 1
       select_type: SIMPLE
             table: student
        partitions: NULL
              type: ref
     possible_keys: sex_index,age_index,nsa_index
               key: age_index
           key_len: 18
               ref: const
              rows: 1
          filtered: 16.67
             Extra: Using where
1 row in set, 1 warning (0.00 sec)

mysql>
```

#按性别索引
```
mysql> explain SELECT sex FROM student WHERE sex='男' \G
*************************** 1. row ***************************
                id: 1
       select_type: SIMPLE
             table: student
        partitions: NULL
              type: ref
     possible_keys: sex_index
               key: sex_index
           key_len: 18
               ref: const
              rows: 3
          filtered: 100.00
             Extra: Using index
1 row in set, 1 warning (0.00 sec)
```

#按照最左前缀匹配原则，它不满足nsa_index联合索引
```
mysql> explain SELECT * FROM student WHERE sex='男' AND age='30'\G
*************************** 1. row ***************************
```

```
          id: 1
  select_type: SIMPLE
        table: student
   partitions: NULL
         type: ref
possible_keys: sex_index,age_index
          key: age_index
      key_len: 18
          ref: const
         rows: 1
     filtered: 50.00
        Extra: Using where
1 row in set, 1 warning (0.00 sec)
```

（3）匹配列前缀索引查询：匹配列的前一部分进行查询，如查询姓名，可以按姓氏"刘"开头进行查询，使用通配符%，但是通配符只能放置在右侧，不能放置在左侧。

```
#通配符放置在右侧
mysql> explain SELECT * FROM student WHERE  name like '刘%' \G
*************************** 1. row ***************************
          id: 1
  select_type: SIMPLE
        table: student
   partitions: NULL
         type: range
possible_keys: nsa_index
          key: nsa_index
      key_len: 768
          ref: NULL
         rows: 1
     filtered: 100.00
        Extra: Using index condition
1 row in set, 1 warning (0.05 sec)

#通配符放置在前面不能使用索引
mysql> explain SELECT * FROM student WHERE  name like '%刘%' \G
*************************** 1. row ***************************
          id: 1
  select_type: SIMPLE
        table: student
   partitions: NULL
         type: ALL
possible_keys: NULL
          key: NULL
      key_len: NULL
          ref: NULL
         rows: 6
     filtered: 16.67
        Extra: Using where
1 row in set, 1 warning (0.04 sec)

mysql>
```

（4）匹配值的范围索引查询：对索引可以进行范围查询。

```
mysql> explain SELECT * FROM student WHERE  id > 2 AND id < 5 \G
*************************** 1. row ***************************
           id: 1
  select_type: SIMPLE
        table: student
   partitions: NULL
         type: range
possible_keys: PRIMARY,id
          key: PRIMARY
      key_len: 4
          ref: NULL
         rows: 2
     filtered: 100.00
        Extra: Using where
1 row in set, 1 warning (0.00 sec)
```

（5）仅对索引进行查询：只针对索引进行查询，也叫索引覆盖，不查询数据表。

例如，在年龄上建立索引后，查询的时候只查询年龄，这样它就可以通过索引树上的年龄值来获取年龄数据。

```
#年龄age建立索引idx_age,查询年龄的时候从索引上直接获取,不需要到数据表获取
mysql> explain SELECT age FROM user WHERE age >20 \G
*************************** 1. row ***************************
           id: 1
  select_type: SIMPLE
        table: user
   partitions: NULL
         type: range
possible_keys: idx_age
          key: idx_age
      key_len: 5
          ref: NULL
         rows: 5
     filtered: 100.00
        Extra: Using where; Using index
1 row in set, 1 warning (0.05 sec)

mysql>
```

7.5.4 索引会用了吗

使用索引的时候，可以匹配全值索引查询、匹配最左前缀索引查询、匹配列全缀索引查询、匹配值的范围索引查询、仅对索引进行查询，那么你真的会用了吗？下面使用学生表student建立一个联合索引nas_index，包括姓名name、年龄age、性别sex、年级grade，删除其他索引，思考一下表7.3中的SQL语句使用到了nas_index（name、age、sex、grade）哪个字段列的索引。

表7.3 SQL查询语句

SQL查询语句	使用哪个索引
SELECT * FROM student	
SELECT * FROM student WHERE name='小明'	
SELECT * FROM student WHERE name='小明' AND sex='男'	

续表

SQL查询语句	使用哪个索引
SELECT * FROM student WHERE sex='男' AND name='小明'	
SELECT * FROM student WHERE name='小明' AND age='30' AND sex='男'	
SELECT * FROM student WHERE name='小明' AND age='30'	
SELECT * FROM student WHERE name='小明' AND age='30' AND sex='男' AND grade='高三'	
SELECT * FROM student WHERE name='小明' AND age='30' AND sex LIKE '男%' AND grade='高三'	
SELECT * FROM student WHERE name='小明' AND age='30' AND sex LIKE '%男%' AND grade='高三'	
SELECT * FROM student WHERE age='30'	
SELECT * FROM student WHERE name='小明' OR sex='男'	
SELECT * FROM student WHERE name='小明' AND sex='男' OR age='30'	
SELECT * FROM student WHERE name='小明' AND (sex='男' OR age='30')	
SELECT * FROM student WHERE name='小明' AND sex='男' AND age > '30' AND grade='高中'	
SELECT * FROM student WHERE name='小明' AND sex='男' AND grade='高中' ORDER BY age	
SELECT * FROM student WHERE name='小明' AND grade='高中' ORDER BY sex,age	
SELECT * FROM student WHERE name='小明' AND grade='高中' GROUP BY sex,age	

实战演练——联合索引应用

（1）把姓名name、性别sex、年龄age、年级grade建立联合索引nsag_index，删除其他索引id、age_index、sex_index、nsa_index。

```
Microsoft Windows [版本 6.1.7601]
版权所有 (c) 2009 Microsoft Corporation。保留所有权利。

C:\Users\Administrator>mysql -uroot -p123456
mysql: [Warning] Using a password on the command line interface can be insecure.
Welcome to the MySQL monitor.  Commands end with ; or \g.
Your MySQL connection id is 3
Server version: 5.7.20 MySQL Community Server (GPL)

Copyright (c) 2000, 2017, Oracle and/or its affiliates. All rights reserved.

Oracle is a registered trademark of Oracle Corporation and/or its
affiliates. Other names may be trademarks of their respective
owners.

Type 'help;' or '\h' for help. Type '\c' to clear the current input statement.

mysql>
mysql> use shop;
Database changed

#查看学生表student的索引情况
mysql> SHOW INDEX FROM student;
+-----+-----+-----+-----+-----+-----+-----+-----+
```

```
| Table   | Non_unique | Key_name  | Seq_in_index | Column_name | Collation | Cardinality | Sub_part | Packed | Null | Index_type | Comment | Index_comment |
```
Table	Non_unique	Key_name	Seq_in_index	Column_name	Collation	Cardinality	Sub_part	Packed	Null	Index_type	Comment	Index_comment
student	0	PRIMARY	1	id	A	6	NULL	NULL		BTREE		
student	0	id	1	id	A	6	NULL	NULL		BTREE		
student	1	sex_index	1	sex	A	2	NULL	NULL	YES	BTREE		
student	1	age_index	1	age	A	4	NULL	NULL	YES	BTREE		
student	1	nsa_index	1	name	A	6	NULL	NULL	YES	BTREE		
student	1	nsa_index	2	sex	A	6	NULL	NULL	YES	BTREE		
student	1	nsa_index	3	age	A	6	NULL	NULL	YES	BTREE		

7 rows in set (0.09 sec)

#删除年龄age_index索引
mysql> DROP index age_index ON student;
Query OK, 0 rows affected (0.26 sec)
Records: 0 Duplicates: 0 Warnings: 0

#删除性别sex_index索引
mysql> DROP index sex_index ON student;
Query OK, 0 rows affected (0.37 sec)
Records: 0 Duplicates: 0 Warnings: 0

#删除id索引
mysql> DROP index id ON student;
Query OK, 0 rows affected (0.50 sec)
Records: 0 Duplicates: 0 Warnings: 0

#删除联合索引nsa_index
mysql> DROP INDEX nsa_index ON student;
Query OK, 0 rows affected (0.31 sec)
Records: 0 Duplicates: 0 Warnings: 0

#添加姓名name、性别sex、年龄age、年级grade的联合索引nsag_index
mysql> ALTER TABLE student ADD INDEX nsag_index(name,sex,age,grade);
Query OK, 0 rows affected (0.33 sec)
Records: 0 Duplicates: 0 Warnings: 0

#查看学生表的索引情况
mysql> SHOW INDEX FROM student;

Table	Non_unique	Key_name	Seq_in_index	Column_name	Collation	Cardinality	Sub_part	Packed	Null	Index_type	Comment	Index_comment
student	0	PRIMARY	1	id	A	6	NULL	NULL		BTREE		
student	1	nsag_index	1	name	A	6	NULL	NULL	YES	BTREE		
student	1	nsag_index	2	sex	A	6	NULL	NULL	YES	BTREE		
student	1	nsag_index	3	age	A	6	NULL	NULL	YES	BTREE		
student	1	nsag_index	4	age	A	6	NULL	NULL	YES	BTREE		

5 rows in set (0.00 sec)

mysql>

（2）SELECT * FROM student，key为NULL，说明没有使用索引。

```
mysql> explain SELECT * FROM student \G
*************************** 1. row ***************************
           id: 1
  select_type: SIMPLE
        table: student
   partitions: NULL
         type: ALL
possible_keys: NULL
          key: NULL
      key_len: NULL
          ref: NULL
         rows: 6
     filtered: 100.00
        Extra: NULL
1 row in set, 1 warning (0.00 sec)

mysql>
```

（3）SELECT * FROM student WHERE name='小明'，按姓名查找，使用联合索引nsag_index，能用到的索引长度是768，说明用到了name列的索引。

```
mysql> explain SELECT * FROM student WHERE name='小明' \G
*************************** 1. row ***************************
           id: 1
  select_type: SIMPLE
        table: student
   partitions: NULL
         type: ref
possible_keys: nsag_index
          key: nsag_index
      key_len: 768
          ref: const
         rows: 1
     filtered: 100.00
        Extra: NULL
1 row in set, 1 warning (0.00 sec)

mysql>
```

（4）SELECT * FROM student WHERE name='小明' AND sex='男'，按姓名和性别查找，使用联合索引nsag_index，能用到的索引长度是786，说明用到了name和sex列的索引。

```
mysql> explain SELECT * FROM student WHERE name='小明' AND sex='男' \G
*************************** 1. row ***************************
           id: 1
  select_type: SIMPLE
        table: student
   partitions: NULL
         type: ref
possible_keys: nsag_index
          key: nsag_index
      key_len: 786
          ref: const,const
         rows: 1
```

```
                 filtered: 100.00
                    Extra: NULL
1 row in set, 1 warning (0.00 sec)

mysql>
```

（5）SELECT * FROM student WHERE sex='男' AND name='小明'，按性别和姓名查找，性别在前姓名在后，使用联合索引nsag_index，能用到的索引长度是786，说明用到了sex和name列的索引，进一步说明sex和name在AND前后没有关系，数据库会自动查找位置。

```
mysql> explain SELECT * FROM student WHERE  sex='男' AND name='小明' \G
*************************** 1. row ***************************
           id: 1
  select_type: SIMPLE
        table: student
   partitions: NULL
         type: ref
possible_keys: nsag_index
          key: nsag_index
      key_len: 786
          ref: const,const
         rows: 1
     filtered: 100.00
        Extra: NULL
1 row in set, 1 warning (0.00 sec)
```

（6）SELECT * FROM student WHERE name='小明' AND age='30' AND sex='男'，按姓名、年龄、性别查找，使用联合索引nsag_index，能用到的索引长度是804，说明使用到了name、age、sex列的索引。

```
mysql> explain SELECT * FROM student WHERE  name='小明' AND age='30' AND sex='男' \G
*************************** 1. row ***************************
           id: 1
  select_type: SIMPLE
        table: student
   partitions: NULL
         type: ref
possible_keys: nsag_index
          key: nsag_index
      key_len: 804
          ref: const,const,const
         rows: 1
     filtered: 100.00
        Extra: NULL
1 row in set, 1 warning (0.00 sec)

mysql>
```

（7）SELECT * FROM student WHERE name='小明' AND age='30'，按姓名、年龄查找，使用联合索引nsag_index，能用到的索引长度是768，它和仅使用到name列的索引长度一致，说明只使用到了name列的索引；联合索引nsag_index 是以name、sex、age、grade的顺序建立的索引，按照最左前缀匹配原则，（name）（name、sex）（name、sex、age）（name、sex、age、grade）这4种是有效的，而（name、age）是无效的，所有智能匹配到（name）。

```
mysql> explain SELECT * FROM student WHERE  name='小明' AND age='30' \G
*************************** 1. row ***************************
```

```
              id: 1
    select_type: SIMPLE
          table: student
     partitions: NULL
           type: ref
  possible_keys: nsag_index
            key: nsag_index
        key_len: 768
            ref: const
           rows: 1
       filtered: 16.67
          Extra: Using index condition
1 row in set, 1 warning (0.00 sec)

mysql>
```

（8）SELECT * FROM student WHERE name='小明' AND age='30' AND sex='男' AND grade='高三'，按姓名、年龄、性别、年级查找，使用联合索引nsag_index，能用到的索引长度是837，说明用到了name、age、sex、grade列的索引。

```
mysql> explain SELECT * FROM student WHERE name='小明' AND age='30' AND sex='男' AND grade='高三'\G
*************************** 1. row ***************************
              id: 1
    select_type: SIMPLE
          table: student
     partitions: NULL
           type: ref
  possible_keys: nsag_index
            key: nsag_index
        key_len: 837
            ref: const,const,const,const
           rows: 1
       filtered: 100.00
          Extra: NULL
1 row in set, 1 warning (0.00 sec)

mysql>
```

（9）SELECT * FROM student WHERE name='小明' AND age='30' AND sex LIKE '男%' AND grade='高三' \G，按姓名、年龄、性别、年级查找，性别采用通配符%匹配列前索引查询，使用联合索引nsag_index，能用到的索引长度是837，说明用到了name、age、sex、grade列的索引。

```
mysql> explain SELECT * FROM student WHERE  name='小明' AND age='30' AND sex LIKE '男%' AND grade='高三' \G
*************************** 1. row ***************************
              id: 1
    select_type: SIMPLE
          table: student
     partitions: NULL
           type: range
  possible_keys: nsag_index
            key: nsag_index
        key_len: 837
            ref: NULL
           rows: 1
```

```
        filtered: 16.67
           Extra: Using index condition
1 row in set, 1 warning (0.00 sec)
```

（10）SELECT * FROM student WHERE name='小明' AND age='30' AND sex LIKE '%男%' AND grade='高三'，按姓名、年龄、性别、年级查找，性别采用通配符%前后进行匹配，使用联合索引nsag_index，能用到的索引长度是768，说明只用到了name列的索引。

```
mysql> explain SELECT * FROM student WHERE  name='小明' AND age='30' AND sex LIKE '%男%' AND grade='高三' \G
*************************** 1. row ***************************
             id: 1
    select_type: SIMPLE
          table: student
     partitions: NULL
           type: ref
  possible_keys: nsag_index
            key: nsag_index
        key_len: 768
            ref: const
           rows: 1
       filtered: 16.67
          Extra: Using index condition
1 row in set, 1 warning (0.00 sec)

mysql>
```

（11）SELECT * FROM student WHERE age='30'，按年龄查找，使用联合索引nsag_index（以name、sex、age、grade的顺序），key为NULL，说明没有用到索引。

```
mysql> explain SELECT * FROM student WHERE  age='30' \G
*************************** 1. row ***************************
             id: 1
    select_type: SIMPLE
          table: student
     partitions: NULL
           type: ALL
  possible_keys: NULL
            key: NULL
        key_len: NULL
            ref: NULL
           rows: 6
       filtered: 16.67
          Extra: Using where
1 row in set, 1 warning (0.00 sec)
```

（12）SELECT * FROM student WHERE name='小明' OR sex='男'，按姓名或者性别查询，使用OR连接时是不能使用索引的，key为NULL，说明没有用到索引。

```
mysql> explain SELECT * FROM student WHERE  name='小明' OR sex='男' \G
*************************** 1. row ***************************
             id: 1
    select_type: SIMPLE
          table: student
     partitions: NULL
           type: ALL
```

```
        possible_keys: nsag_index
                  key: NULL
              key_len: NULL
                  ref: NULL
                 rows: 6
             filtered: 30.56
                Extra: Using where
1 row in set, 1 warning (0.00 sec)
```

（13）SELECT * FROM student WHERE name='小明' AND sex='男' OR age='30'，按姓名、性别或者年龄查询，OR连接的是不能使用索引的，它相当于（name='小明' AND sex='男'）OR age='30'， key为NULL，说明没有用到索引。

```
mysql> explain SELECT * FROM student WHERE  name='小明' AND sex='男' OR age='30' \G
*************************** 1. row ***************************
           id: 1
  select_type: SIMPLE
        table: student
   partitions: NULL
         type: ALL
possible_keys: nsag_index
          key: NULL
      key_len: NULL
          ref: NULL
         rows: 6
     filtered: 18.98
        Extra: Using where
1 row in set, 1 warning (0.00 sec)
```

（14）SELECT * FROM student WHERE name='小明' AND（sex='男' OR age='30'），按姓名、性别或者年龄查询，OR连接的是不能使用索引的，但是AND前面的name是可以使用索引的，使用联合索引nsag_index，能用到的索引长度是768，说明只用到了name列的索引。

```
mysql> explain SELECT * FROM student WHERE  name='小明' AND (sex='男' OR age='30') \G
*************************** 1. row ***************************
           id: 1
  select_type: SIMPLE
        table: student
   partitions: NULL
         type: ref
possible_keys: nsag_index
          key: nsag_index
      key_len: 768
          ref: const
         rows: 1
     filtered: 30.56
        Extra: Using index condition
1 row in set, 1 warning (0.00 sec)
```

（15）SELECT * FROM student WHERE name='小明' AND sex='男' AND age > '30' AND grade='高中'，按姓名、性别、年龄、年级查询，年龄采用范围查询，使用联合索引nsag_index，能用到的索引长度是804，说明使用到了name、sex、age列的索引。

```
mysql> explain SELECT * FROM student WHERE  name='小明' AND sex='男' AND age > '30' AND grade='高中' \G
*************************** 1. row ***************************
           id: 1
```

```
        select_type: SIMPLE
              table: student
         partitions: NULL
               type: range
      possible_keys: nsag_index
                key: nsag_index
            key_len: 804
                ref: NULL
               rows: 1
           filtered: 16.67
              Extra: Using index condition
1 row in set, 1 warning (0.00 sec)
```

（16）SELECT * FROM student WHERE name='小明' AND sex='男' AND grade='高中' ORDER BY age，按姓名、性别、年级查询，按年龄排序，使用联合索引nsag_index，能用到的索引长度是786，说明用到了name和sex列的索引。

```
mysql> explain SELECT * FROM student WHERE name='小明' AND sex='男' AND grade='高中' ORDER BY age \G
*************************** 1. row ***************************
                 id: 1
        select_type: SIMPLE
              table: student
         partitions: NULL
               type: ref
      possible_keys: nsag_index
                key: nsag_index
            key_len: 786
                ref: const,const
               rows: 1
           filtered: 16.67
              Extra: Using index condition
1 row in set, 1 warning (0.00 sec)
```

（17）SELECT * FROM student WHERE name='小明' AND grade='高中' ORDER BY sex,age，按姓名、年级查询，按性别、年龄排序，使用联合索引nsag_index，能用到的索引长度是768，说明用到了name列的索引。

```
mysql> explain SELECT * FROM student WHERE name='小明' AND grade='高中' ORDER BY sex,age \G
*************************** 1. row ***************************
                 id: 1
        select_type: SIMPLE
              table: student
         partitions: NULL
               type: ref
      possible_keys: nsag_index
                key: nsag_index
            key_len: 768
                ref: const
               rows: 1
           filtered: 16.67
              Extra: Using index condition
1 row in set, 1 warning (0.00 sec)

mysql>
```

（18）SELECT * FROM student WHERE name='小明' AND grade='高中' GROUP BY sex,age，按

姓名、年级查询，按性别、年龄分组，使用联合索引nsag_index，能用到的索引长度是768，说明用到了name列的索引。

```
mysql> explain SELECT * FROM student WHERE name='小明' AND grade='高中' GROUP BY sex,age \G
*************************** 1. row ***************************
           id: 1
  select_type: SIMPLE
        table: student
   partitions: NULL
         type: ref
possible_keys: nsag_index
          key: nsag_index
      key_len: 768
          ref: const
         rows: 1
     filtered: 16.67
        Extra: Using index condition
1 row in set, 1 warning (0.00 sec)
```

下面回答一下表7.3中SQL查询语句用到哪些列的索引，建立的联合索引nsag_index(name,sex,age,grade)，到底使用到联合索引的哪一列呢？如表7.4所示。

表7.4　SQL查询语句索引使用情况

SQL查询语句	使用哪个索引
SELECT * FROM student	无索引
SELECT * FROM student WHERE name='小明'	name列索引
SELECT * FROM student WHERE name='小明' AND sex='男'	name、sex列索引
SELECT * FROM student WHERE sex='男' AND name='小明'	name、sex列索引
SELECT * FROM student WHERE name='小明' AND age='30' AND sex='男'	name、age、sex列索引
SELECT * FROM student WHERE name='小明' AND age='30'	name列索引
SELECT * FROM student WHERE name='小明' AND age='30' AND sex='男' AND grade='高三'	name、age、sex、grade列索引
SELECT * FROM student WHERE name='小明' AND age='30' AND sex LIKE '男%' AND grade='高三'	name、age、sex、grade列索引
SELECT * FROM student WHERE name='小明' AND age='30' AND sex LIKE '%男%' AND grade='高三'	name列索引
SELECT * FROM student WHERE age='30'	无索引
SELECT * FROM student WHERE name='小明' OR sex='男'	无索引
SELECT * FROM student WHERE name='小明' AND sex='男' OR age='30'	无索引
SELECT * FROM student WHERE name='小明' AND (sex='男' OR age='30')	name列索引
SELECT * FROM student WHERE name='小明' AND sex='男' AND age > '30' AND grade='高中'	name、sex、age列索引
SELECT * FROM student WHERE name='小明' AND sex='男' AND grade='高中' ORDER BY age	name、sex列索引
SELECT * FROM student WHERE name='小明' AND grade='高中' ORDER BY sex,age	name列索引
SELECT * FROM student WHERE name='小明' AND grade='高中' GROUP BY sex,age	name列索引

7.6 索引的类型

7.6.1 主键索引

数据库在建表的时候，都会建立表的主键，这时就可以根据表的主键建立索引，称为主键索引（PRIMARY KEY）。主键索引是不允许重复和不允许有空值的，它是唯一索引的一种特例。下面建立一个课程表course，有主键id、课程的名称，根据主键id建立主键索引。

精讲视频

索引的类型

实战演练——主键索引

```
Microsoft Windows [版本 6.1.7601]
版权所有 (c) 2009 Microsoft Corporation。保留所有权利。

C:\Users\Administrator>mysql -uroot -p123456
mysql: [Warning] Using a password on the command line interface can be insecure.
Welcome to the MySQL monitor.  Commands end with ; or \g.
Your MySQL connection id is 7
Server version: 5.7.20 MySQL Community Server (GPL)

Copyright (c) 2000, 2017, Oracle and/or its affiliates. All rights reserved.

Oracle is a registered trademark of Oracle Corporation and/or its
affiliates. Other names may be trademarks of their respective
owners.

Type 'help;' or '\h' for help. Type '\c' to clear the current input statement.

mysql> USE shop
Database changed

#创建课程表course，并创建主键索引primary key
mysql> CREATE TABLE course(
    -> id int not null AUTO_INCREMENT,
    -> name varchar(255),
    -> primary key(id)
    -> );
Query OK, 0 rows affected (0.38 sec)

#查看索引，可以看到主键索引创建成功
mysql> SHOW INDEX FROM course;
+--------+------------+----------+--------------+-------------+-----------+-------------+----------+--------+------+------------+---------+---------------+
| Table  | Non_unique | Key_name | Seq_in_index | Column_name | Collation | Cardinality | Sub_part | Packed | Null | Index_type | Comment | Index_comment |
+--------+------------+----------+--------------+-------------+-----------+-------------+----------+--------+------+------------+---------+---------------+
| course |          0 | PRIMARY  |            1 | id          | A         |           0 |     NULL | NULL   |      | BTREE      |         |               |
+--------+------------+----------+--------------+-------------+-----------+-------------+----------+--------+------+------------+---------+---------------+
1 row in set (0.00 sec)

mysql>
```

7.6.2 普通索引

普通索引是最基本的索引，没有任何限制，可以在创建表的时候创建索引，可以创建表后直接创建索引，也可以以修改表结构的方式添加索引。下面创建一个成绩表score，有主键id、学号num、姓名name、分数grade，在创建表时将学号添加索引，创建表后通过直接创建索引的方式将姓名添加索引，通过修改成绩表score的表结构添加索引。

实战演练 —— 普通索引

```
Microsoft Windows [版本 6.1.7601]
版权所有 (c) 2009 Microsoft Corporation。保留所有权利。

C:\Users\Administrator>mysql –uroot –p123456
mysql: [Warning] Using a password on the command line interface can be insecure.
Welcome to the MySQL monitor.  Commands end with ; or \g.
Your MySQL connection id is 8
Server version: 5.7.20 MySQL Community Server (GPL)

Copyright (c) 2000, 2017, Oracle and/or its affiliates. All rights reserved.

Oracle is a registered trademark of Oracle Corporation and/or its
affiliates. Other names may be trademarks of their respective
owners.

Type 'help;' or '\h' for help. Type '\c' to clear the current input statement.

mysql> use shop
Database changed

#创建成绩表，在创建表的时候将学号num添加索引
mysql> CREATE TABLE score(
    -> id int not null AUTO_INCREMENT,
    -> num int not null,
    -> name varchar(255),
    -> grade varchar(20),
    -> primary key(id),
    -> INDEX num_index(num)
    -> );
Query OK, 0 rows affected (0.41 sec)

#直接以创建索引的方式将姓名name添加索引
mysql> CREATE INDEX name_index ON score(name);
Query OK, 0 rows affected (0.21 sec)
Records: 0  Duplicates: 0  Warnings: 0

#以修改成绩表score的表结构的方式添加索引
mysql> ALTER TABLE score ADD INDEX grade_index(grade);
Query OK, 0 rows affected (0.26 sec)
Records: 0  Duplicates: 0  Warnings: 0

#查看score表的索引
mysql> SHOW INDEX FROM score;
```

```
+-------+------------+------------+--------------+-------------+-----------+-------------+----------+--------+------+------------+---------+---------------+
| Table | Non_unique | Key_name   | Seq_in_index | Column_name | Collation | Cardinality | Sub_part | Packed | Null | Index_type | Comment | Index_comment |
+-------+------------+------------+--------------+-------------+-----------+-------------+----------+--------+------+------------+---------+---------------+
| score |      0     | PRIMARY    |       1      |     id      |     A     |      0      |   NULL   |  NULL  |      |   BTREE    |         |               |
| score |      1     | num_index  |       1      |    num      |     A     |      0      |   NULL   |  NULL  |      |   BTREE    |         |               |
| score |      1     | name_index |       1      |    name     |     A     |      0      |   NULL   |  NULL  |   Y  |   BTREE    |         |               |
| score |      1     | grade_index|       1      |    grade    |     A     |      0      |   NULL   |  NULL  |   Y  |   BTREE    |         |               |
+-------+------------+------------+--------------+-------------+-----------+-------------+----------+--------+------+------------+---------+---------------+
4 rows in set (0.00 sec)

mysql>
```

7.6.3 唯一索引

唯一索引要求列值不允许重复，但是允许有空值，主键索引就是唯一索引的特例。也可以通过联合索引来创建唯一索引，要求列值的组合必须唯一，需要使用关键字UNIQUE来标识唯一索引。下面创建一个成绩表score，该表有主键id、学号num、姓名name、分数grade，在创建表时将学号添加唯一索引，创建表后通过直接创建索引的方式将姓名添加唯一索引，通过修改成绩表score的表结构添加唯一索引。

实战演练 ——唯一索引

```
Microsoft Windows [版本 6.1.7601]
版权所有 (c) 2009 Microsoft Corporation。保留所有权利。

C:\Users\Administrator>mysql -uroot -p123456
mysql: [Warning] Using a password on the command line interface can be insecure.
Welcome to the MySQL monitor.  Commands end with ; or \g.
Your MySQL connection id is 8
Server version: 5.7.20 MySQL Community Server (GPL)

Copyright (c) 2000, 2017, Oracle and/or its affiliates. All rights reserved.

Oracle is a registered trademark of Oracle Corporation and/or its
affiliates. Other names may be trademarks of their respective
owners.

Type 'help;' or '\h' for help. Type '\c' to clear the current input statement.

mysql> use shop
Database changed

#创建成绩表，在创建表的时候将学号num添加唯一索引
mysql> CREATE TABLE score(
    -> id int not null AUTO_INCREMENT,
    -> num int not null,
    -> name varchar(255),
    -> grade varchar(20),
    -> primary key(id),
    -> UNIQUE INDEX num_index(num)
    -> );
```

Query OK, 0 rows affected (0.41 sec)

#通过直接创建索引的方式将姓名name添加唯一索引
```
mysql> CREATE UNIQUE INDEX name_index ON score(name);
Query OK, 0 rows affected (0.21 sec)
Records: 0  Duplicates: 0  Warnings: 0
```

#通过修改成绩表score的表结构添加唯一索引
```
mysql> ALTER TABLE score ADD UNIQUE INDEX grade_index(grade);
Query OK, 0 rows affected (0.26 sec)
Records: 0  Duplicates: 0  Warnings: 0
```

#查看score表的索引
```
mysql> SHOW INDEX FROM score;
+-------+------------+-------------+--------------+-------------+-----------+-------------+----------+--------+------+------------+---------+---------------+
| Table | Non_unique | Key_name    | Seq_in_index | Column_name | Collation | Cardinality | Sub_part | Packed | Null | Index_type | Comment | Index_comment |
+-------+------------+-------------+--------------+-------------+-----------+-------------+----------+--------+------+------------+---------+---------------+
| score |          0 | PRIMARY     |            1 | id          | A         |           0 |     NULL | NULL   |      | BTREE      |         |               |
| score |          0 | num_index   |            1 | num         | A         |           0 |     NULL | NULL   |      | BTREE      |         |               |
| score |          0 | name_index  |            1 | name        | A         |           0 |     NULL | NULL   |      | BTREE      |         |               |
| score |          0 | grade_index |            1 | grade       | A         |           0 |     NULL | NULL   |      | BTREE      |         |               |
+-------+------------+-------------+--------------+-------------+-----------+-------------+----------+--------+------+------------+---------+---------------+
4 rows in set (0.00 sec)

mysql>
```

7.6.4 单列索引和联合索引

单列索引就是一个索引包含单个列，一个表可以有多个单列索引；而联合索引是一个索引包含多个列。下面创建一个成绩表score，该表有主键id、学号num、姓名name、分数grade，在创建表时将学号添加单列索引，创建表后通过直接创建索引的方式将学号、姓名、分数创建联合索引。

实战演练——单列索引和联合索引

```
Microsoft Windows [版本 6.1.7601]
版权所有 (c) 2009 Microsoft Corporation。保留所有权利。

C:\Users\Administrator>mysql -uroot -p123456
mysql: [Warning] Using a password on the command line interface can be insecure.
Welcome to the MySQL monitor.  Commands end with ; or \g.
Your MySQL connection id is 8
Server version: 5.7.20 MySQL Community Server (GPL)

Copyright (c) 2000, 2017, Oracle and/or its affiliates. All rights reserved.

Oracle is a registered trademark of Oracle Corporation and/or its
affiliates. Other names may be trademarks of their respective
owners.

Type 'help;' or '\h' for help. Type '\c' to clear the current input statement.
```

```
mysql> use shop
Database changed

#创建成绩表, 在创建表的时候将学号num添加单列索引
mysql> CREATE TABLE score(
    -> id int not null AUTO_INCREMENT,
    -> num int not null,
    -> name varchar(255),
    -> grade varchar(20),
    -> primary key(id),
    -> INDEX num_index(num)
    -> );
Query OK, 0 rows affected (0.34 sec)

#将学号、姓名、分数建立联合索引
mysql> CREATE INDEX nng_index ON score(num,name,grade);
Query OK, 0 rows affected (0.23 sec)
Records: 0  Duplicates: 0  Warnings: 0

#查看索引
mysql> SHOW INDEX FROM score;
+-------+------------+-----------+--------------+-------------+-----------+-------------+----------+--------+------+------------+---------+---------------+
| Table | Non_unique | Key_name  | Seq_in_index | Column_name | Collation | Cardinality | Sub_part | Packed | Null | Index_type | Comment | Index_comment |
+-------+------------+-----------+--------------+-------------+-----------+-------------+----------+--------+------+------------+---------+---------------+
| score |          0 | PRIMARY   |            1 | id          | A         |           0 |     NULL | NULL   |      | BTREE      |         |               |
| score |          0 | nng_index |            1 | num         | A         |           1 |     NULL | NULL   |      | BTREE      |         |               |
| score |          0 | nng_index |            1 | name        | A         |           2 |     NULL | NULL   |      | BTREE      |         |               |
| score |          0 | nng_index |            1 | grade       | A         |           3 |     NULL | NULL   |      | BTREE      |         |               |
| score |          1 | num_index |            1 | num         | A         |           0 |     NULL | NULL   |      | BTREE      |         |               |
+-------+------------+-----------+--------------+-------------+-----------+-------------+----------+--------+------+------------+---------+---------------+
5 rows in set (0.00 sec)
```

7.6.5 聚簇索引和非聚簇索引

聚簇索引是将数据存放在索引树的叶子节点上,找到叶子节点就可以读取这行数据。InnoDB存储引擎的索引方式就是聚簇索引。一个表只能有一个聚簇索引,一般会根据主键或者唯一索引,或者以数据库内部生成的rowid为主键,来建立聚簇索引。

非聚簇索引是在索引树的叶子节点上存放数据的地址,找到该地址后,需要到磁盘中查询一次才能获取到数据。MyISAM存储引擎的索引方式就是非聚餐索引,只在索引树的叶子节点上存放地址。

7.6.6 覆盖索引

覆盖索引是指查询的列正好是索引的一部分,那么它直接从索引上获取数据,而不需要到磁盘中查找数据,这种查询效率非常高。下面创建一个成绩表score,该表有主键id、学号num、姓名name、分数grade,在创建表时将学号添加索引,然后查询学号。

```
Microsoft Windows [版本 6.1.7601]
版权所有 (c) 2009 Microsoft Corporation。保留所有权利。

C:\Users\Administrator>mysql –uroot –p123456
```

```
mysql: [Warning] Using a password on the command line interface can be insecure.
Welcome to the MySQL monitor.  Commands end with ; or \g.
Your MySQL connection id is 8
Server version: 5.7.20 MySQL Community Server (GPL)

Copyright (c) 2000, 2017, Oracle and/or its affiliates. All rights reserved.

Oracle is a registered trademark of Oracle Corporation and/or its
affiliates. Other names may be trademarks of their respective
owners.

Type 'help;' or '\h' for help. Type '\c' to clear the current input statement.

mysql> use shop
Database changed

#创建成绩表,在创建表的时候将学号num添加单列索引
mysql> CREATE TABLE score(
    -> id int not null AUTO_INCREMENT,
    -> num int not null,
    -> name varchar(255),
    -> grade varchar(20),
    -> primary key(id),
    -> INDEX num_index(num)
    -> );
Query OK, 0 rows affected (0.34 sec)

#查询学号num,直接从索引树获取num数据
mysql> explain SELECT num FROM score WHERE num = 10 \G
*************************** 1. row ***************************
           id: 1
  select_type: SIMPLE
        table: score
   partitions: NULL
         type: ref
possible_keys: num_index
          key: num_index
      key_len: 4
          ref: const
         rows: 1
     filtered: 100.00
        Extra: Using index
1 row in set, 1 warning (0.00 sec)
```

7.6.7 重复索引和冗余索引

重复索引是指在同一列上按照相同的顺序创建同类型的索引,如在年龄age这一列上,创建了(age)单列索引,又创建了(age,sex)联合索引,按照最左前缀匹配原则,联合索引(age,sex)包含了(age)单列索引,所以(age)单列索引就是重复的,应该删除。

冗余索引是指如果创建了联合索引(age,sex),再创建(age)索引,就是冗余索引,但是创建(sex,age)就不是冗余索引,因为顺序不同,索引效果也就不同,碰到冗余索引应该删除。

未使用索引是指除了冗余索引和重复索引，可能还会有一些MySQL数据库永远不使用的索引，这样的索引完全是累赘，建议删除。

MySQL允许在同一列上创建多个索引，这些索引都需要单独维护，会增加维护成本，并且优化器在优化查询的时候也需要逐个地考虑这些索引，会影响性能，所有重复索引、冗余索引、未使用索引都需要考虑删除。

7.7 索引不能使用

7.7.1 不等于（<>、!=）

使用不等于（<>、!=）操作符进行条件判断时是不能使用索引的，如在学生表student中有姓名name、性别sex、年龄age单列索引，下面查找姓名name不等于小明的数据，查看索引的使用情况。

精讲视频

索引不能使用

实战演练 ——不等于不能使用索引

Microsoft Windows [版本 6.1.7601]
版权所有 (c) 2009 Microsoft Corporation。保留所有权利。

C:\Users\Administrator>mysql –uroot –p123456
mysql: [Warning] Using a password on the command line interface can be insecure.
Welcome to the MySQL monitor. Commands end with ; or \g.
Your MySQL connection id is 9
Server version: 5.7.20 MySQL Community Server (GPL)

Copyright (c) 2000, 2017, Oracle and/or its affiliates. All rights reserved.

Oracle is a registered trademark of Oracle Corporation and/or its
affiliates. Other names may be trademarks of their respective
owners.

Type 'help;' or '\h' for help. Type '\c' to clear the current input statement.

mysql> USE shop;
Database changed

#查看数据库表
mysql> SELECT * FROM student;
+----+--------+-----+-----+-------+-------+------+
| id | name | sex | age | grade | class | num |
+----+--------+-----+-----+-------+-------+------+
1	刘明	男	19	高三	6班	3001
2	吴倩	女	18	高三	6班	3002
3	张欣	男	17	高二	1班	2001
4	孙晓	女	17	高二	1班	2002
5	赵英俊	男	16	高一	2班	1001
6	柳师师	女	17	高一	2班	1002
+----+--------+-----+-----+-------+-------+------+
6 rows in set (0.00 sec)

#查看学生表student

```
mysql> SHOW INDEX FROM student;
+---------+------------+------------+--------------+-------------+-----------+-------------+----------+--------+------+------------+---------+---------------+
| Table   | Non_unique | Key_name   | Seq_in_index | Column_name | Collation | Cardinality | Sub_part | Packed | Null | Index_type | Comment | Index_comment |
+---------+------------+------------+--------------+-------------+-----------+-------------+----------+--------+------+------------+---------+---------------+
| student | 0          | PRIMARY    | 1            | id          | A         | 1           | NULL     | NULL   |      | BTREE      |         |               |
| student | 0          | id         | 1            | id          | A         | 1           | NULL     | NULL   |      | BTREE      |         |               |
| student | 1          | nsa_index  | 1            | name        | A         | 1           | NULL     | NULL   | YES  | BTREE      |         |               |
| student | 1          | nsa_index  | 1            | sex         | A         | 2           | NULL     | NULL   | YES  | BTREE      |         |               |
| student | 1          | nsa_index  | 1            | age         | A         | 3           | NULL     | NULL   | YES  | BTREE      |         |               |
| student | 1          | nsa_index  | 1            | grade       | A         | 4           | NULL     | NULL   | YES  | BTREE      |         |               |
+---------+------------+------------+--------------+-------------+-----------+-------------+----------+--------+------+------------+---------+---------------+
6 rows in set (0.00 sec)

#删除联合索引
mysql> DROP INDEX nsag_index ON student;
Query OK, 0 rows affected (0.25 sec)
Records: 0  Duplicates: 0  Warnings: 0

#创建单列索引|姓名name
mysql> CREATE INDEX name_index ON student(name);
Query OK, 0 rows affected (0.27 sec)
Records: 0  Duplicates: 0  Warnings: 0

#创建单列索引|性别sex
mysql> CREATE INDEX sex_index ON student(sex);
Query OK, 0 rows affected (0.26 sec)
Records: 0  Duplicates: 0  Warnings: 0

#创建单列索引|年龄age
mysql> CREATE INDEX age_index ON student(age);
Query OK, 0 rows affected (0.50 sec)
Records: 0  Duplicates: 0  Warnings: 0

#按姓名列索引|查询姓名等于小明的数据，使用到姓名索引|name_index
mysql> explain SELECT * FROM student WHERE name = '小明' \G
*************************** 1. row ***************************
           id: 1
  select_type: SIMPLE
        table: student
   partitions: NULL
         type: ref
possible_keys: name_index
          key: name_index
      key_len: 768
          ref: const
         rows: 1
     filtered: 100.00
        Extra: NULL
1 row in set, 1 warning (0.00 sec)

#按姓名列索引|查询姓名不等于（<>）小明的数据，没有使用到索引
```

```
mysql> explain SELECT * FROM student WHERE name <> '小明' \G
*************************** 1. row ***************************
           id: 1
  select_type: SIMPLE
        table: student
   partitions: NULL
         type: ALL
possible_keys: name_index
          key: NULL
      key_len: NULL
          ref: NULL
         rows: 6
     filtered: 100.00
        Extra: Using where
1 row in set, 1 warning (0.00 sec)

#按姓名列索引查询姓名不等于（!=）小明的数据，没有使用到索引
mysql> explain SELECT * FROM student WHERE name != '小明' \G
*************************** 1. row ***************************
           id: 1
  select_type: SIMPLE
        table: student
   partitions: NULL
         type: ALL
possible_keys: name_index
          key: NULL
      key_len: NULL
          ref: NULL
         rows: 6
     filtered: 100.00
        Extra: Using where
1 row in set, 1 warning (0.00 sec)

mysql>
```

7.7.2 前导模糊查询（%xx%）

使用前导模糊查询（%xx%）不能使用索引，但是后导模糊查询（xx%）是可以使用索引的。下面在学生表student的姓名name列添加索引，然后按姓名name进行前导模糊查询和后导模糊查询，查看索引的使用情况。

实战演练——前导模糊查询不能使用索引

```
Microsoft Windows [版本 6.1.7601]
版权所有 (c) 2009 Microsoft Corporation。保留所有权利。

C:\Users\Administrator>mysql -uroot -p123456
mysql: [Warning] Using a password on the command line interface can be insecure.

Welcome to the MySQL monitor. Commands end with ; or \g.
Your MySQL connection id is 10
Server version: 5.7.20 MySQL Community Server (GPL)
```

```
Copyright (c) 2000, 2017, Oracle and/or its affiliates. All rights reserved.

Oracle is a registered trademark of Oracle Corporation and/or its
affiliates. Other names may be trademarks of their respective
owners.

Type 'help;' or '\h' for help. Type '\c' to clear the current input statement.

mysql> USE shop;
Database changed
```

#前导模糊查询（%xx%），没有使用到索引
```
mysql> explain SELECT * FROM student WHERE name like '%小明%' \G
*************************** 1. row ***************************
           id: 1
  select_type: SIMPLE
        table: student
   partitions: NULL
         type: ALL
possible_keys: NULL
          key: NULL
      key_len: NULL
          ref: NULL
         rows: 6
     filtered: 16.67
        Extra: Using where
1 row in set, 1 warning (0.00 sec)
```

#后导模糊查询（xx%），使用姓名name列索引
```
mysql> explain SELECT * FROM student WHERE name like '小明%' \G
*************************** 1. row ***************************
           id: 1
  select_type: SIMPLE
        table: student
   partitions: NULL
         type: range
possible_keys: name_index
          key: name_index
      key_len: 768
          ref: NULL
         rows: 1
     filtered: 100.00
        Extra: Using index condition
1 row in set, 1 warning (0.00 sec)

mysql>
```

7.7.3 比较不匹配的数据类型

比较不匹配的数据类型是指某一列在比较查询时，假如列是varchar字符串类型，但是在赋值的时候赋值int整型，这时这一列即使有索引，也不会使用到索引。

实战演练 ——比较不匹配的数据类型不能使用索引

```
Microsoft Windows [版本 6.1.7601]
版权所有 (c) 2009 Microsoft Corporation。保留所有权利。

C:\Users\Administrator>mysql -uroot -p123456
mysql: [Warning] Using a password on the command line interface can be insecure.

Welcome to the MySQL monitor.  Commands end with ; or \g.
Your MySQL connection id is 11
Server version: 5.7.20 MySQL Community Server (GPL)

Copyright (c) 2000, 2017, Oracle and/or its affiliates. All rights reserved.

Oracle is a registered trademark of Oracle Corporation and/or its
affiliates. Other names may be trademarks of their respective
owners.

Type 'help;' or '\h' for help. Type '\c' to clear the current input statement.

mysql> USE shop
Database changed

#姓名name列是字符串类型，赋值字符串类型
mysql> SELECT * FROM student WHERE name='刘明';
+----+------+-----+-----+-------+-------+------+
| id | name | sex | age | grade | class | num  |
+----+------+-----+-----+-------+-------+------+
|  1 | 刘明 | 男  | 19  | 高三  | 6班   | 3001 |
+----+------+-----+-----+-------+-------+------+
1 row in set (0.00 sec)

#姓名name列是字符串类型，赋值整型
mysql> SELECT * FROM student WHERE name=1;
Empty set, 6 warnings (0.04 sec)

#姓名name列是字符串类型，赋值字符串类型，可以使用姓名name列索引
mysql> explain  SELECT * FROM student WHERE name='小明' \G
*************************** 1. row ***************************
           id: 1
  select_type: SIMPLE
        table: student
   partitions: NULL
         type: ref
possible_keys: name_index
          key: name_index
      key_len: 768
          ref: const
         rows: 1
     filtered: 100.00
        Extra: NULL
```

```
1 row in set, 1 warning (0.00 sec)
```

#姓名name列是字符串类型，赋值整型，类型不匹配，不可以使用name列索引
```
mysql> explain  SELECT * FROM student WHERE name=1 \G
*************************** 1. row ***************************
           id: 1
  select_type: SIMPLE
        table: student
   partitions: NULL
         type: ALL
possible_keys: name_index
          key: NULL
      key_len: NULL
          ref: NULL
         rows: 6
     filtered: 16.67
        Extra: Using where
1 row in set, 3 warnings (0.00 sec)

mysql>
```

7.7.4　OR连接条件

在SQL语句的条件中，用OR连接的条件表达式不能使用索引。在写SQL语句的时候，应尽量不使用OR，因为这样会导致其他条件表达式即使加了索引也不会起作用，从而降低查询效率。

实战演练——OR连接条件

```
Microsoft Windows [版本 6.1.7601]
版权所有 (c) 2009 Microsoft Corporation。保留所有权利。

C:\Users\Administrator>mysql -uroot -p123456
mysql: [Warning] Using a password on the command line interface can be insecure.

Welcome to the MySQL monitor.  Commands end with ; or \g.
Your MySQL connection id is 11
Server version: 5.7.20 MySQL Community Server (GPL)

Copyright (c) 2000, 2017, Oracle and/or its affiliates. All rights reserved.

Oracle is a registered trademark of Oracle Corporation and/or its
affiliates. Other names may be trademarks of their respective
owners.

Type 'help;' or '\h' for help. Type '\c' to clear the current input statement.

mysql> USE shop
Database changed
```

#使用AND连接条件表达式，可以使用索引
```
mysql> explain SELECT * FROM student WHERE name = '小明' AND sex='男' \G
*************************** 1. row ***************************
```

```
            id: 1
  select_type: SIMPLE
        table: student
   partitions: NULL
         type: ref
possible_keys: name_index,sex_index
          key: name_index
      key_len: 768
          ref: const
         rows: 1
     filtered: 50.00
        Extra: Using where
1 row in set, 1 warning (0.00 sec)
```

#使用OR连接条件表达式，不能使用索引
```
mysql> explain SELECT * FROM student WHERE name = '小明' OR sex='男' \G
*************************** 1. row ***************************
            id: 1
  select_type: SIMPLE
        table: student
   partitions: NULL
         type: ALL
possible_keys: name_index,sex_index
          key: NULL
      key_len: NULL
          ref: NULL
         rows: 6
     filtered: 30.56
        Extra: Using where
1 row in set, 1 warning (0.00 sec)
```

#使用OR连接条件表达式，不能使用索引
```
mysql> explain SELECT * FROM student WHERE name = '小明' AND sex='男' OR age='20' \G
*************************** 1. row ***************************
            id: 1
  select_type: SIMPLE
        table: student
   partitions: NULL
         type: ALL
possible_keys: name_index,sex_index,age_index
          key: NULL
      key_len: NULL
          ref: NULL
         rows: 6
     filtered: 18.98
        Extra: Using where
1 row in set, 1 warning (0.00 sec)
```

#使用AND连接和OR连接的两个表达式放置在括号里，可以使用索引
```
mysql> explain SELECT * FROM student WHERE name = '小明' AND (sex='男' OR age='20') \G
*************************** 1. row ***************************
            id: 1
```

```
     select_type: SIMPLE
            table: student
       partitions: NULL
             type: ref
    possible_keys: name_index,sex_index,age_index
              key: name_index
          key_len: 768
              ref: const
             rows: 1
         filtered: 30.56
            Extra: Using where
1 row in set, 1 warning (0.00 sec)

mysql>
```

7.7.5 条件表达式前使用函数

在SQL语句的条件中，在表达式前使用函数或者运算操作，不能使用索引；在表达式后使用函数或者运算操作，可以使用索引。

实战演练 ——条件表达式前使用函数

```
Microsoft Windows [版本 6.1.7601]
版权所有 (c) 2009 Microsoft Corporation。保留所有权利。

C:\Users\Administrator>mysql –uroot –p123456
mysql: [Warning] Using a password on the command line interface can be insecure.

Welcome to the MySQL monitor.  Commands end with ; or \g.
Your MySQL connection id is 11
Server version: 5.7.20 MySQL Community Server (GPL)

Copyright (c) 2000, 2017, Oracle and/or its affiliates. All rights reserved.

Oracle is a registered trademark of Oracle Corporation and/or its
affiliates. Other names may be trademarks of their respective
owners.

Type 'help;' or '\h' for help. Type '\c' to clear the current input statement.

mysql> USE shop
Database changed

#使用id进行查询，id列是添加索引的，可以使用索引
mysql> explain SELECT * FROM student WHERE id = 1 \G
*************************** 1. row ***************************
               id: 1
      select_type: SIMPLE
            table: student
       partitions: NULL
             type: const
    possible_keys: PRIMARY
```

```
          key: PRIMARY
      key_len: 4
          ref: const
         rows: 1
     filtered: 100.00
        Extra: NULL
1 row in set, 1 warning (0.00 sec)
```

#在条件表达式前进行运算操作，不能使用索引
```
mysql> explain SELECT * FROM student WHERE id+1 = 1 \G
*************************** 1. row ***************************
           id: 1
  select_type: SIMPLE
        table: student
   partitions: NULL
         type: ALL
possible_keys: NULL
          key: NULL
      key_len: NULL
          ref: NULL
         rows: 6
     filtered: 100.00
        Extra: Using where
1 row in set, 1 warning (0.02 sec)
```

#在表达式后进行运算操作，可以使用索引
```
mysql> explain SELECT * FROM student WHERE id = 1 + 1 \G
*************************** 1. row ***************************
           id: 1
  select_type: SIMPLE
        table: student
   partitions: NULL
         type: const
possible_keys: PRIMARY
          key: PRIMARY
      key_len: 4
          ref: const
         rows: 1
     filtered: 100.00
        Extra: NULL
1 row in set, 1 warning (0.00 sec)
```

#在表达式后使用函数可以使用索引，在表达式前使用函数不能使用索引
```
mysql> explain SELECT * FROM student WHERE name = concat('姓名','小明') \G
*************************** 1. row ***************************
           id: 1
  select_type: SIMPLE
        table: student
   partitions: NULL
         type: ref
possible_keys: name_index
          key: name_index
```

```
                key_len: 768
                    ref: const
                   rows: 1
               filtered: 100.00
                  Extra: NULL
1 row in set, 1 warning (0.00 sec)
```

7.8 索引的利弊及建立原则

索引可以提高查询效率、排序以及分组，但在效率提高的同时，也会带来一些负面影响。那么，索引有哪些优势和劣势呢？建立索引的原则又有哪些呢？

精讲视频

索引的利弊及建立原则

索引的优势如下。

（1）索引最大的一个优势就是提高查询效率，它可以通过创建唯一索引或者主键索引来标识行的唯一性，在查询的时候可以快速定位到要查询的行数据。

（2）可以加快表与表之间的连接查询。

（3）在分组和排序的时候可以极大地节省时间，使用索引可以进行快速排序，分组查询时虽然不能直接使用索引，但是分组查询要先进行排序，而在排序这个阶段就会减少时间，所以分组查询也能节省时间。

（4）使用索引来进行查询、排序、分组，使用优化隐藏器提高系统的性能。

索引的劣势如下。

（1）创建索引和维护索引需要耗费时间，随着数据量的增大时间也会逐渐增加。

（2）索引文件的大小会逐渐增大，索引文件需要占用物理空间，如果建立聚簇索引，文件大小会更大，会使数据库的存储逐渐变大。

（3）对表进行增加、修改、删除的时候，都需要对索引进行维护，这会影响对表和数据的操作速度，延长对表的操作时间。

索引的建立原则如下。

（1）主键的字段需要建立唯一索引或者主键索引。

（2）用于连接查询的字段（外键）可以建立索引。

（3）经常用来排序的字段可以建立索引。

（4）频繁使用WHERE的条件字段可以建立索引。

（5）唯一性太差的字段不适合建立索引。

（6）更新频率远大于查询的字段不适合建立索引。

（7）不会出现WHERE条件表达式的字段不适合建立索引。

（8）对于类型为text、blob、image、bit的字段，不适合建立索引。

7.9 小结

索引对于数据库查询、排序、分组是非常重要的，它可以提高数据库查询、排序、分组的效率。掌握索引的使用，有助于解决SQL语句优化，提高数据库的查询性能。通过本章的学习要学会怎么创建索引、查询索引、删除索引；学会索引的数据结构，知道什么是B-Tree数据结构、B+Tree数据结构；学会常见存储引擎的索引实现原理，包括MyISAM的索引实现、InnoDB的索引实现、MEMORY的索引实现；学会应用索引、使用explain分析索引以及索引的使用策略；了解索引的类型及其使用；了解索引在哪些情况下不能使用；了解索引的优势、劣势和建立索引的原则。

第8章
综合案例——图书管理系统

本章要点

- 需求管理
- 数据库设计
- 创建数据库
- 用户信息管理
- 图书管理
- 借书管理
- 视图管理
- 小结

■ 图书管理系统能使借书人借书更加方便,同时减轻图书管理员的负担,使借书、还书操作更加系统化、规范化,同时方便图书管理,实现图书查询、图书借阅、图书归还、图书管理、图书统计等功能。

8.1 需求管理

图书管理系统有三类角色：普通用户、图书管理员和系统管理员。这三类角色的需求如下。

（1）普通用户：查看个人信息、修改个人信息、查询图书、借阅图书。

（2）图书管理员：图书预约查询、图书借阅查询、借阅图书登记、还书登记、图书遗失登记。

（3）系统管理员：管理用户、分配角色权限、系统设置。

业务流程如下：

（1）普通用户来到图书管理员处借书，告知图书管理员是否有预约。如果有预约，可以直接登记借出图书；如果没有预约，告知图书管理员相关书籍信息，管理员再进行查询、登记等操作。

（2）普通用户来到图书管理员处还书，告知图书管理员书籍情况，然后图书管理员进行还书登记。

精讲视频

需求管理

8.2 数据库设计

图书管理系统是围绕用户和图书进行操作的系统，需要建立图书管理系统数据库books，建立用户表user、部门表dept、角色表role、图书表book、图书分类表book_classify、图书借阅表book_borrow、图书还书表book_return、借阅预约表book_appoint、图书遗失表book_lose。各表包含的信息如下。

（1）用户表：包含用户编号、姓名、出生日期、身份证号、登录名称、登录密码、手机号、电子邮箱、部门编号、角色编号信息，如表8.1所示。

精讲视频

数据库设计

表8.1 用户表user

字　　段	字 段 名 称	字 段 类 型	备　　注
id	编号	int	主键
user_name	姓名	varchar(255)	
birth_date	出生日期	date	yyyy-MM-dd
id_card	身份证号	varchar(255)	
login_name	登录名称	varchar(255)	
password	登录密码	varchar(255)	
mobile	手机号	varchar(255)	
email	电子邮箱	varchar(255)	
dept_id	部门编号	int	与部门表关联
role_id	角色编号	int	与角色表关联，0：普通用户、1：图书管理员、2：系统管理员

（2）部门表：包含部门编号、部门名称、创建日期信息，如表8.2所示。

表8.2 部门表dept

字　　段	字 段 名 称	字 段 类 型	备　　注
id	部门编号	Int	主键
dept_name	部门名称	varchar(255)	
create_date	创建日期	Date	yyyy-MM-dd

（3）角色表：包含角色编号、角色名称、备注信息，如表8.3所示。

表8.3　角色表role

字　　段	字　段　名　称	字　段　类　型	备　　注
id	角色编号	int	主键
role_name	角色名称	varchar(255)	
remark	备注	varchar(255)	

（4）图书表：包含图书编号、图书名称、作者、图书定价、是否有光盘、出版社、图书分类编号、图书总数量、图书ISBN编号、图书创建时间、备注信息，如表8.4所示。

表8.4　图书表book

字　　段	字　段　名　称	字　段　类　型	备　　注
id	图书编号	int	主键
book_name	图书名称	varchar(255)	
author	作者	varchar(255)	
price	图书定价	decimal	
cd	是否有光盘	int	0：有、1：无
publish	出版社	varchar(50)	出版社
book_classify_id	图书分类编号	int	与图书分类表关联
account	图书总数量	int	
isbn	图书ISBN编号	varchar(50)	
create_time	图书创建时间	datetime	yyyy-mm-dd HH:mm:ss
remark	备注	varchar(255)	

（5）图书分类表：包含编号、图书分类名称、父分类编号、创建时间信息，如表8.5所示。

表8.5　图书分类表book_classify

字　　段	字　段　名　称	字　段　类　型	备　　注
id	编号	int	主键
book_classify_name	图书分类名称	varchar(255)	
father_id	父分类编号	int	顶级父类编号为0
create_time	创建时间	datetime	yyyy-mm-dd HH:mm:ss

（6）图书借阅表：包含图书借阅编号、图书编号、用户编号、借阅时间、归还时间、创建时间、备注信息，如表8.6所示。

表8.6　图书借阅表book_borrow

字　　段	字　段　名　称	字　段　类　型	备　　注
id	图书借阅编号	int	主键
book_id	图书编号	int	
user_id	用户编号	int	顶级父类编号为0
borrow_time	借阅时间	date	yyyy-mm-dd

续表

字 段	字 段 名 称	字 段 类 型	备 注
return_time	归还时间	date	yyyy-mm-dd
create_time	创建时间	datetime	yyyy-mm-dd HH:mm:ss
remark	备注	varchar(255)	

（7）图书还书表：包含还书流水编号、图书借阅编号、归还时间、创建时间、备注信息，如表8.7所示。

表8.7　图书还书表book_return

字 段	字 段 名 称	字 段 类 型	备 注
id	还书流水编号	int	主键
borrow_id	图书借阅编号	int	
return_time	归还时间	date	yyyy-mm-dd
create_time	创建时间	datetime	yyyy-mm-dd HH:mm:ss
remark	备注	varchar(255)	

（8）借阅预约表：包含预约流水编号、图书编号、用户编号、预约时间、创建时间、备注信息，如表8.8所示。

表8.8　借阅预约表book_appoint

字 段	字 段 名 称	字 段 类 型	备 注
id	预约流水编号	int	主键
book_id	图书编号	int	
user_id	用户编号	int	
appoint_time	预约时间	date	yyyy-mm-dd
create_time	创建时间	datetime	yyyy-mm-dd HH:mm:ss
remark	备注	varchar(255)	

（9）图书遗失表：包含遗失流水编号、图书借阅流水编号、登记时间、备注信息，如表8.9所示。

表8.9　图书遗失表book_lose

字 段	字 段 名 称	字 段 类 型	备 注
id	遗失流水编号	int	主键
borrow_id	图书借阅流水编号	int	
create_time	登记时间	datetime	yyyy-mm-dd HH:mm:ss
remark	备注	varchar(255)	

8.3　创建数据库

在图书管理系统数据库books中，建立用户表user、部门表dept、角色表role、图书表book、图书分类表book_classify、图书借阅表book_borrow、图书还书表book_return、借阅预约表book_appoint和图书遗失表book_lose。

精讲视频

创建数据库

8.3.1 建表语句

（1）创建图书管理系统数据库books。

create database books;

（2）创建用户表user。

```
DROP TABLE IF EXISTS `user`;
CREATE TABLE `user` (
  `id` int(11) NOT NULL,
  `user_name` varchar(255) DEFAULT NULL,
  `birth_date` date DEFAULT NULL,
  `id_card` varchar(255) DEFAULT NULL,
  `login_name` varchar(255) DEFAULT NULL,
  `password` varchar(255) DEFAULT NULL,
  `mobile` varchar(255) DEFAULT NULL,
  `email` varchar(255) DEFAULT NULL,
  `dept_id` int(11) DEFAULT NULL,
  `role_id` int(11) DEFAULT NULL,
  PRIMARY KEY (`id`)
) ENGINE=InnoDB DEFAULT CHARSET=utf8;
```

（3）创建部门表dept。

```
DROP TABLE IF EXISTS `dept`;
CREATE TABLE `dept` (
  `id` int(11) NOT NULL,
  `dept_name` varchar(255) DEFAULT NULL,
  `create_date` date DEFAULT NULL,
  PRIMARY KEY (`id`)
) ENGINE=InnoDB DEFAULT CHARSET=utf8;
```

（4）创建角色表role。

```
DROP TABLE IF EXISTS `role`;
CREATE TABLE `role` (
  `id` int(11) NOT NULL,
  `role_name` varchar(255) DEFAULT NULL,
  `remark` varchar(255) DEFAULT NULL,
  PRIMARY KEY (`id`)
) ENGINE=InnoDB DEFAULT CHARSET=utf8;
```

（5）创建图书表book。

```
DROP TABLE IF EXISTS `book`;
CREATE TABLE `book` (
  `id` int(11) NOT NULL,
  `book_name` varchar(255) DEFAULT NULL,
  `author` varchar(255) DEFAULT NULL,
  `price` decimal(10,0) DEFAULT NULL,
  `cd` int(11) DEFAULT NULL,
  `publish` varchar(50) DEFAULT NULL,
  `book_classify_id` int(11) DEFAULT NULL,
  `account` int(11) DEFAULT NULL,
  `isbn` varchar(50) DEFAULT NULL,
  `create_time` datetime DEFAULT NULL,
  `remark` varchar(255) DEFAULT NULL,
  PRIMARY KEY (`id`)
) ENGINE=InnoDB DEFAULT CHARSET=utf8;
```

（6）创建图书分类表book_classify。

```sql
DROP TABLE IF EXISTS `book_classify`;
CREATE TABLE `book_classify` (
  `id` int(11) NOT NULL,
  `book_classify_name` varchar(255) DEFAULT NULL,
  `father_id` int(11) DEFAULT NULL,
  `create_time` datetime DEFAULT NULL,
  PRIMARY KEY (`id`)
) ENGINE=InnoDB DEFAULT CHARSET=utf8;
```

（7）创建图书借阅表book_borrow。

```sql
DROP TABLE IF EXISTS `book_borrow`;
CREATE TABLE `book_borrow` (
  `id` int(11) NOT NULL,
  `book_id` int(11) DEFAULT NULL,
  `user_id` int(11) DEFAULT NULL,
  `borrow_time` date DEFAULT NULL,
  `return_time` date DEFAULT NULL,
  `create_time` datetime DEFAULT NULL,
  `remark` varchar(255) DEFAULT NULL,
  PRIMARY KEY (`id`)
) ENGINE=InnoDB DEFAULT CHARSET=utf8;
```

（8）创建还书表book_return。

```sql
DROP TABLE IF EXISTS `book_return`;
CREATE TABLE `book_return` (
  `id` int(11) NOT NULL,
  `borrow_id` int(11) DEFAULT NULL,
  `return_time` date DEFAULT NULL,
  `create_time` datetime DEFAULT NULL,
  `remark` varchar(255) DEFAULT NULL,
  PRIMARY KEY (`id`)
) ENGINE=InnoDB DEFAULT CHARSET=utf8;
```

（9）创建图书预约表book_appoint。

```sql
DROP TABLE IF EXISTS `book_appoint`;
CREATE TABLE `book_appoint` (
  `id` int(11) NOT NULL,
  `book_id` int(11) DEFAULT NULL,
  `user_id` int(11) DEFAULT NULL,
  `appoint_time` date DEFAULT NULL,
  `create_time` datetime DEFAULT NULL,
  `remark` varchar(255) DEFAULT NULL,
  PRIMARY KEY (`id`)
) ENGINE=InnoDB DEFAULT CHARSET=utf8;
```

（10）创建图书遗失表book_lose。

```sql
DROP TABLE IF EXISTS `book_lose`;
CREATE TABLE `book_lose` (
  `id` int(11) NOT NULL,
  `borrow_id` int(11) DEFAULT NULL,
  `create_time` datetime DEFAULT NULL,
  `remark` varchar(255) DEFAULT NULL,
  PRIMARY KEY (`id`)
) ENGINE=InnoDB DEFAULT CHARSET=utf8;
```

8.3.2 初始化数据

（1）插入数据到用户表user。

INSERT INTO `user` VALUES ('1', '小刚', '1985-05-25', '230822111122223333', 'xiaogang', '123456', '15211112222', null, '1', '1');

INSERT INTO `user` VALUES ('2', '小影', '1989-02-19', '10121111122223333', 'xiaoying', '123456', '13811112222', null, '1', '1');

INSERT INTO `user` VALUES ('3', '大梅', '1989-06-01', '237881111122223333', 'damei', '123456', '13811112222', null, '2', '2');

INSERT INTO `user` VALUES ('4', 'admin', '1988-03-25', '234567111122223333', 'admin', '123456', '15122223333', null, '3', '3');

（2）插入数据到角色表role。

INSERT INTO `role` VALUES ('1', '普通用户', null);
INSERT INTO `role` VALUES ('2', '图书管理员', null);
INSERT INTO `role` VALUES ('3', '系统管理员', null);

（3）插入数据到部门表dept。

INSERT INTO `dept` VALUES ('1', '行政部', '2018-06-03');
INSERT INTO `dept` VALUES ('2', '后勤部', '2018-06-03');
INSERT INTO `dept` VALUES ('3', '运维部', '2018-06-03');

（4）插入数据到图书表book。

INSERT INTO `book` VALUES ('1', '微信小程序开发图解案例教程', '刘刚', '59.80', '1', '人民邮电出版社', '3', '1000', '9787115450456', '2018-06-10 14:20:56', null);

INSERT INTO `book` VALUES ('2', 'Axure RP8原型设计图解视频教程Web+App', '刘刚', '79.80', '1', '人民邮电出版社', '3', '2000', '9787115445131', '2018-06-03 14:26:36', null);

INSERT INTO `book` VALUES ('3', '军事小百科（儿童仿真版）（套装共6册）[6-12岁]', '张柏赫，李京键', '76.80', '1', '吉林出版集团股份有限公司', '1', '500', '9787553478203', '2018-06-03 14:28:52', null);

INSERT INTO `book` VALUES ('4', '牛奶可乐经济学', '[美] 罗伯特·弗兰克 著，闾佳 译', '49.90', '1', '北京联合出版公司', '2', '500', '9787550292505', '2018-06-03 14:34:17', null);

（5）插入数据到图书借阅表book_borrow。

INSERT INTO `book_borrow` VALUES ('1', '1', '2', '2018-04-01', '2018-06-03', '2018-04-01 14:35:29', null);

INSERT INTO `book_borrow` VALUES ('2', '2', '1', '2018-05-14', '2018-07-14', '2018-05-14 14:36:37', null);

INSERT INTO `book_borrow` VALUES ('3', '4', '2', '2018-05-01', '2018-05-30', '2018-05-01 14:41:18', null);

（6）插入数据到图书分类表book_classify。

INSERT INTO `book_classify` VALUES ('1', 'E 军事', '0', '2018-06-03 14:23:05');
INSERT INTO `book_classify` VALUES ('2', 'F 经济', '0', '2018-06-03 14:23:35');
INSERT INTO `book_classify` VALUES ('3', 'T 工业技术', '0', '2018-06-03 14:24:16');

（7）插入数据到图书预约表book_appoint。

INSERT INTO `book_appoint` VALUES ('1', '1', '2', '2018-03-30', '2018-03-30 14:37:40', null);
INSERT INTO `book_appoint` VALUES ('2', '2', '1', '2018-05-10', '2018-05-10 14:38:31', null);

（8）插入数据到还书表book_return。

INSERT INTO `book_return` VALUES ('1', '1', '2018-06-03', '2018-06-03 14:39:30', null);
INSERT INTO `book_return` VALUES ('2', '2', '2018-06-14', '2018-06-14 14:40:12', null);

（9）插入数据到图书遗失表book_lose。

INSERT INTO `book_lose` VALUES ('1', '3', '2018-05-30 14:41:38', null);

8.4 用户信息管理

用户信息管理包括对用户、部门的管理。用户管理包括新增用户、删除用户、修改用户；部门管理包括新增部门、修改部门、删除部门。

8.4.1 用户管理

我们用以下案例来简要介绍用户管理的部分功能。

（1）新增一个借阅图书的用户小红，隶属于行政部，角色是普通用户。

```
INSERT INTO `user` VALUES ('5', '小红', '1985-06-27', '1012011111122223333', 'xiaohong', '123456', '15211112222', null, '1', '1');
```

（2）修改用户小红的部门，修改成运维部。

```
UPDATE user SET dept_id = 2 WHERE login_name='xiaohong';
```

（3）查询用户小红的姓名、登录名、部门名称、角色名称。

```
SELECT u.user_name,u.login_name,d.dept_name,r.role_name FROM user u, dept d, role r WHERE u.dept_id = d.id AND u.role_id = r.id AND u.login_name = 'xiaohong';
```

输出结果：

```
+-----------+------------+-----------+-----------+
| user_name | login_name | dept_name | role_name |
+-----------+------------+-----------+-----------+
| 小红      | xiaohong   | 后勤部    | 普通用户  |
+-----------+------------+-----------+-----------+
```

（4）删除用户小红。

```
DELETE FROM user WHERE login_name = 'xiaohong';
```

8.4.2 部门管理

我们用以下案例来简要介绍部门管理的部分功能。

（1）新增一个办公室部门。

```
INSERT INTO `dept` VALUES ('4', '办公室', '2018-06-08');
```

（2）查询所有部门。

```
SELECT * FROM dept;;
```

输出结果：

```
+----+-----------+-------------+
| id | dept_name | create_date |
+----+-----------+-------------+
| 1  | 行政部    | 2018-06-03  |
| 2  | 后勤部    | 2018-06-03  |
| 3  | 运维部    | 2018-06-03  |
| 4  | 办公室    | 2018-06-08  |
+----+-----------+-------------+
```

8.5 图书管理

图书管理即图书分类管理，可以新增图书分类、修改图书分类；可以新增图书以及修改图书信息。

8.5.1 新增图书分类

我们用以下案例来简要介绍新增图书分类的部分功能。

在图书分类里面，新增一类图书分类R医药、卫生，把分类信息插入到图书分类表book_classify中。
INSERT INTO `book_classify` VALUES ('4', 'R 医药、卫生', '0', '2018-06-08 14:23:05');
查询所有的图书分类信息。
SELECT * FROM book_classify;
查询结果：

```
+----+-------------------+-----------+---------------------+
| id | book_classify_name| father_id | create_time         |
+----+-------------------+-----------+---------------------+
|  1 | E 军事            |         0 | 2018-06-03 14:23:05 |
|  2 | F 经济            |         0 | 2018-06-03 14:23:35 |
|  3 | T 工业技术        |         0 | 2018-06-03 14:24:16 |
|  4 | R 医药、卫生      |         0 | 2018-06-08 14:23:05 |
+----+-------------------+-----------+---------------------+
```

8.5.2 新增图书

我们用以下案例来简要介绍新增图书的部分功能。

图书管理系统里需要新增一本刚采购的医药方面的书籍：书名为《疑难杂病临证手册（第2版）》、作者为余孟学、定价为158元、出版社为河南科技出版社、ISBN编号为9787534989230。要把它添加到图书表book里，图书分类选择"R 医药、卫生"。

INSERT INTO `book` VALUES ('5', '疑难杂病临证手册（第2版）', '余孟学', '158', '1', '河南科技出版社', '4', '1000', '9787534989230', '2018-06-22 18:20:56', null);
查询所有图书的图书名称、作者、图书定价、出版社、图书分类名称、图书总数量。
SELECT b.book_name,b.author,b.price,b.publish,c.book_classify_name,b.account FROM book b,book_classify c WHERE b.book_classify_id = c.id;
查询结果：

8.6 借书管理

借书管理是图书管理系统的核心，可以进行借书预约管理、借书登记管理、还书管理、图书遗失登记管理，通过对图书的借阅登记、还书登记、图书遗失登记的管理，来达到对图书的管理。

8.6.1 借书预约管理

用户小影要借一本《Axure RP8原型设计图解视频教程Web+App》，在借书前，需要先通过图书管理系统进行借书预约登记。

精讲视频

借书管理

(1)借书预约登记。
INSERT INTO `book_appoint` VALUES ('1', '2', '2', '2018-04-30', '2018-04-30 14:37:40', null);
(2)查询用户小影的借书预约记录,包括用户姓名、借阅图书名称、作者、图书总数量、预约登记时间。
SELECT u.user_name,b.book_name,w.borrow_time,l.create_time FROM user u,book b,book_borrow w,book_lose l WHERE w.book_id=b.id AND w.user_id=u.id AND w.id = l.borrow_id;

查询结果:

```
+---------+-----------------------------------------+--------+---------+--------------+
| user_name | book_name                             | author | account | appoint_time |
+---------+-----------------------------------------+--------+---------+--------------+
| 小影    | 微信小程序开发图解案例教程              | 刘刚   | 1000    | 2018-03-30   |
| 小影    | Axure RP8原型设计图解视频教程Web+App    | 刘刚   | 2000    | 2018-04-30   |
+---------+-----------------------------------------+--------+---------+--------------+
```

8.6.2 借书登记管理

用户小影要借一本《Axure RP8原型设计图解视频教程Web+App》,要进行借书登记。

(1)借书登记。
INSERT INTO `book_borrow` VALUES ('4', '2', '2', '2018-06-01', '2018-08-01', '2018-06-01 14:35:29', null);
(2)查询用户小影的借书记录,包括用户姓名、借阅图书名称、出版社、借书时间、归还时间。
SELECT u.user_name,b.book_name,b.publish,w.borrow_time,w.return_time from user u, book b, book_borrow w where w.book_id = b.id AND w.user_id = u.id AND u.login_name = 'xiaoying';

查询结果:

```
+---------+------------------------------------------+----------------+-------------+-------------+
| user_name | book_name                              | publish        | borrow_time | return_time |
+---------+------------------------------------------+----------------+-------------+-------------+
| 小影    | 微信小程序开发图解案例教程               | 人民邮电出版社 | 2018-04-01  | 2018-06-03  |
| 小影    | 牛奶可乐经济学                           | 北京联合出版公司| 2018-05-01 | 2018-05-30  |
| 小影    | Axure RP8原型设计图解视频教程Web+App     | 人民邮电出版社 | 2018-06-01  | 2018-08-01  |
+---------+------------------------------------------+----------------+-------------+-------------+
```

8.6.3 还书管理

用户小影要归还《Axure RP8原型设计图解视频教程Web+App》,需要进行还书登记。

(1)还书登记。
INSERT INTO `book_return` VALUES ('3', '4', '2018-07-30', '2018-07-30 14:40:12', null);
(2)查询用户小影的还书记录,包括用户姓名、借阅图书名称、借书时间、还书时间。
SELECT u.user_name,b.book_name,w.borrow_time,r.return_time FROM user u,book b,book_borrow w,book_return r WHERE w.book_id = b.id AND w.user_id = u.id AND w.id = r.borrow_id AND u.login_name = 'xiaoying';

查询结果:

```
+---------+-----------------------------------------+-------------+-------------+
| user_name | book_name                             | borrow_time | return_time |
+---------+-----------------------------------------+-------------+-------------+
| 小影    | 微信小程序开发图解案例教程              | 2018-04-01  | 2018-06-03  |
| 小影    | Axure RP8原型设计图解视频教程Web+App    | 2018-06-01  | 2018-07-30  |
+---------+-----------------------------------------+-------------+-------------+
```

8.6.4 图书遗失登记管理

用户小刚借阅的图书《Axure RP8原型设计图解视频教程Web+App》遗失了,对遗失的图书要进行图书遗失登记。

（1）图书遗失登记。
INSERT INTO `book_lose` VALUES ('3', '2', '2018-08-01 14:41:38', null);
（2）查询图书遗失记录，包括用户姓名、借阅书籍名称、借阅时间、遗失登记时间。
SELECT u.user_name,b.book_name,b.author,b.account,a.appoint_time FROM user u,book b,book_appoint a WHERE a.book_id = b.id AND a.user_id = u.id AND u.login_name = 'xiaoying';
查询结果：

```
+-----------+-----------------------------------------------+-------------+---------------------+
| user_name | book_name                                     | borrow_time | create_time         |
+-----------+-----------------------------------------------+-------------+---------------------+
| 小影      | 牛奶可乐经济学                                | 2018-05-01  | 2018-05-30 14:41:38 |
| 小刚      | Axure RP8原型设计图解视频教程Web+App          | 2018-05-14  | 2018-08-01 14:41:38 |
+-----------+-----------------------------------------------+-------------+---------------------+
```

8.7 视图管理

对于多表连接查询时，可以对多个表建立视图，然后从视图里进行查询，这样可以提高查询速度。我们可以以用户信息查询、用户借阅图书查询、用户还书查询建立视图。

8.7.1 用户信息查询视图

针对用户表user、部门表dept、角色表role建立一个用户信息查询视图user_info_view，查询用户编号、姓名、登录名称、部门名称、角色名称。
CREATE OR REPLACE VIEW user_info_view
AS
SELECT u.id,u.user_name,u.login_name,d.dept_name,r.role_name
FROM user u, dept d, role r
WHERE u.dept_id = d.id AND u.role_id = r.id ;
视图查询结果：

```
+----+-----------+------------+-----------+--------------+
| id | user_name | login_name | dept_name | role_name    |
+----+-----------+------------+-----------+--------------+
| 1  | 小刚      | xiaogang   | 行政部    | 普通用户     |
| 2  | 小影      | xiaoying   | 行政部    | 普通用户     |
| 3  | 大梅      | damei      | 后勤部    | 图书管理员   |
| 4  | admin     | admin      | 运维部    | 系统管理员   |
+----+-----------+------------+-----------+--------------+
```

8.7.2 用户借阅图书查询视图

针对用户表user、图书表book、图书借阅表book_borrow建立一个用户借阅图书信息查询视图user_book_borrow_view，查询用户编号、登录名称、姓名、图书名称、出版社、借阅时间、归还时间。
CREATE OR REPLACE VIEW user_book_borrow_view
AS
SELECT u.id,u.login_name,u.user_name,b.book_name,b.publish,w.borrow_time,w.return_time
from user u, book b, book_borrow w
where w.book_id = b.id AND w.user_id = u.id ;
视图查询结果：

```
+----+------------+-----------+-----------+---------+-------------+-------------+
| id | login_name | user_name | book_name | publish | borrow_time | return_time |
```

```
+----+------------+-----------+-------------------+---------------------+------------+------------+
|  2 | xiaoying   | 小影      | 牛奶可乐经济学    | 北京联合出版公司    | 2018-05-01 | 2018-05-30 |
+----+------------+-----------+-------------------+---------------------+------------+------------+
```

8.7.3 用户还书查询视图

针对用户表user、图书表book、图书借阅表book_borrow、还书表book_return建立一个用户还书查询视图user_book_return_view，查询用户编号、登录名称、姓名、图书名称、借阅时间、归还时间。

```
CREATE OR REPLACE VIEW user_book_return_view
AS
SELECT u.id,u.login_name,u.user_name,b.book_name,w.borrow_time,r.return_time
FROM user u,book b,book_borrow w,book_return r
WHERE w.book_id = b.id AND w.user_id = u.id AND w.id = r.borrow_id ;
```

视图查询结果：

```
+----+------------+-----------+----------------------------------+-------------+-------------+
| id | login_name | user_name | book_name                        | borrow_time | return_time |
+----+------------+-----------+----------------------------------+-------------+-------------+
|  2 | xiaoying   | 小影      | 微信小程序开发图解案例教程       | 2018-04-01  | 2018-06-03  |
+----+------------+-----------+----------------------------------+-------------+-------------+
```

8.8 小结

本章练习图书管理系统数据库的设计和使用，建立9张数据库表，包括用户表、部门表、图书表、图书借阅表等；然后插入初始化数据；接着进行用户信息管理、图书管理、借书管理等操作；对于多表联合查询，通过建立视图的方式加快数据库的查询。

第9章
MySQL管理

本章要点

用户管理
权限管理
表空间管理（InnoDB）
备份与还原
主从同步配置
小结

■ 本章介绍MySQL管理，包括用户管理、权限管理、表空间管理、备份与还原，以及主从同步配置。

9.1 用户管理

MySQL数据库的用户管理包括创建用户、修改用户、删除用户等操作。

9.1.1 创建用户

```
#创建用户，任何ip的客户端都可以访问
create user 'xiaogang'@'%' identified by '123456';

#创建用户，只有本地的客户端才可以访问
create user 'xiaogang'@'localhost' identified by '123456';

#创建用户，只有指定的192.168.1.90这个ip才可以访问
create user 'xiaogang'@'192.168.1.90' identified by '123456';
```

用户管理

MySQL数据库使用create关键字来创建用户，在@符号前面是用户的名称，在@符号后面是限制访问数据库客户端的ip。

（1）使用%，则说明任何ip的客户端都可以连接到这个数据库上进行相关操作。

（2）使用localhost，则说明只能允许数据库服务器的本地客户端连接到这个数据库上进行相关操作。

（3）使用指定的ip，如192.168.1.90，则只有这个ip的客户端才可以访问，为了安全起见，往往会设置指定的ip来连接数据库。

采用关键字identified by来指定密码，可以根据自己的需要设置密码的复杂度。创建的用户是允许重复的，如果要查询数据库有哪些用户以及访问数据库的客户端ip权限，则可以使用以下命令。

```
SELECT user,host FROM mysql.user;
```

下面创建一个用户名为xiaogang的用户，密码设置为123456，然后查询数据库里有哪些用户。

```
Microsoft Windows [版本 6.1.7601]
版权所有 (c) 2009 Microsoft Corporation。保留所有权利。

C:\Users\Administrator>mysql -uroot -p123456
mysql: [Warning] Using a password on the command line interface can be insecure.

Welcome to the MySQL monitor.  Commands end with ; or \g.
Your MySQL connection id is 3
Server version: 5.7.20 MySQL Community Server (GPL)

Copyright (c) 2000, 2017, Oracle and/or its affiliates. All rights reserved.

Oracle is a registered trademark of Oracle Corporation and/or its
affiliates. Other names may be trademarks of their respective
owners.

Type 'help;' or '\h' for help. Type '\c' to clear the current input statement.

#创建xiaogang用户，任何ip的客户端都可以访问
mysql> create user 'xiaogang'@'%' identified by '123456';
Query OK, 0 rows affected (0.22 sec)

#创建xiaoying用户，只有本地的客户端才可以访问
mysql> create user 'xiaoying'@'localhost' identified by '123456';
Query OK, 0 rows affected (0.00 sec)
```

```
#创建xiaoming用户，只有指定的192.168.1.90这个ip才可以访问
mysql> create user 'xiaoming'@'192.168.1.90' identified by '123456';
Query OK, 0 rows affected (0.00 sec)

#查询数据库有哪些用户
mysql> SELECT user,host FROM mysql.user;
+---------------+--------------+
| user          | host         |
+---------------+--------------+
| shopdb        | %            |
| xiaogang      | %            |
| shopdb2       | 10.120.71.89 |
| xiaoming      | 192.168.1.90 |
| mysql.session | localhost    |
| mysql.sys     | localhost    |
| root          | localhost    |
| xiaoying      | localhost    |
+---------------+--------------+
8 rows in set (0.00 sec)
```

9.1.3 删除用户

使用新创建的用户连接数据库，如下所示。

```
Microsoft Windows [版本 6.1.7601]
版权所有 (c) 2009 Microsoft Corporation。保留所有权利。

#使用xiaogang用户连接数据库
C:\Users\Administrator>mysql -uxiaogang -p123456
mysql: [Warning] Using a password on the command line interface can be insecure.

Welcome to the MySQL monitor.  Commands end with ; or \g.
Your MySQL connection id is 4
Server version: 5.7.20 MySQL Community Server (GPL)

Copyright (c) 2000, 2017, Oracle and/or its affiliates. All rights reserved.

Oracle is a registered trademark of Oracle Corporation and/or its
affiliates. Other names may be trademarks of their respective
owners.

Type 'help;' or '\h' for help. Type '\c' to clear the current input statement.

mysql>
```

9.1.2 修改用户

MySQL 5.7版本的数据库采用以下命令来修改用户的密码。

```
update mysql.user set authentication_string=password(新密码) where user='xiaogang' and host='localhost';
```

用户客户端访问数据库的ip限制，可以采用以下命令进行修改。

```
update mysql.user set host='192.168.1.100' where user='xiaoying';
```

创建好的用户如果要修改密码，要更新MySQL的用户表，在5.7版本的MySQL中，密码存放在authentication_string中，需要对它进行修改。在修改密码的时候，用户需要获得reload权限，否则使用flush privileges刷新MySQL的系统权限相关表会报错，可以使用以下命令进行授权。

```
GRANT reload ON *.* to 'root'@'%';
```

如果不对用户授予reload权限，也可以在更新密码之后重启MySQL服务，不使用flush privileges刷新MySQL的系统权限相关表。

下面将用户为xiaogang的密码设置为123456789，并将xiaoying的客户端ip限制改为192.168.1.100，如下所示。

```
Microsoft Windows [版本 6.1.7601]
版权所有 (c) 2009 Microsoft Corporation。保留所有权利。

#使用root用户
C:\Users\Administrator>mysql -uroot -p123456
mysql: [Warning] Using a password on the command line interface can be insecure.

Welcome to the MySQL monitor.  Commands end with ; or \g.
Your MySQL connection id is 14
Server version: 5.7.20 MySQL Community Server (GPL)

Copyright (c) 2000, 2017, Oracle and/or its affiliates. All rights reserved.

Oracle is a registered trademark of Oracle Corporation and/or its
affiliates. Other names may be trademarks of their respective
owners.

Type 'help;' or '\h' for help. Type '\c' to clear the current input statement.

#修改用户xiaogang的密码为123456789
mysql> update mysql.user set authentication_string=password('123456789') where user='xiaogang' and host='%';
Query OK, 1 row affected, 1 warning (0.00 sec)
Rows matched: 1  Changed: 1  Warnings: 1

#退出连接
mysql> exit
Bye

#使用xiaogang用户和修改后的密码进行连接，连接失败
C:\Users\Administrator>mysql -uxiaogang -p123456789
mysql: [Warning] Using a password on the command line interface can be insecure.

ERROR 1045 (28000): Access denied for user 'xiaogang'@'localhost' (using password: YES)

#使用root用户
C:\Users\Administrator>mysql -uroot -p123456
mysql: [Warning] Using a password on the command line interface can be insecure.

Welcome to the MySQL monitor.  Commands end with ; or \g.
Your MySQL connection id is 16
Server version: 5.7.20 MySQL Community Server (GPL)

Copyright (c) 2000, 2017, Oracle and/or its affiliates. All rights reserved.

Oracle is a registered trademark of Oracle Corporation and/or its
```

affiliates. Other names may be trademarks of their respective
owners.

Type 'help;' or '\h' for help. Type '\c' to clear the current input statement.

#已经拥有reload权限，如果没有需要授权
mysql> flush privileges;
Query OK, 0 rows affected (0.00 sec)

#退出连接
mysql> exit
Bye

#使用xiaogang用户和修改后的密码进行连接，连接成功
C:\Users\Administrator>mysql –uxiaogang –p123456789
mysql: [Warning] Using a password on the command line interface can be insecure.

Welcome to the MySQL monitor. Commands end with ; or \g.
Your MySQL connection id is 17
Server version: 5.7.20 MySQL Community Server (GPL)

Copyright (c) 2000, 2017, Oracle and/or its affiliates. All rights reserved.

Oracle is a registered trademark of Oracle Corporation and/or its
affiliates. Other names may be trademarks of their respective
owners.

Type 'help;' or '\h' for help. Type '\c' to clear the current input statement.

mysql>

#退出连接
mysql> exit
Bye

#使用root用户连接
C:\Users\Administrator>mysql –uroot –p123456
mysql: [Warning] Using a password on the command line interface can be insecure.

Welcome to the MySQL monitor. Commands end with ; or \g.
Your MySQL connection id is 18
Server version: 5.7.20 MySQL Community Server (GPL)

Copyright (c) 2000, 2017, Oracle and/or its affiliates. All rights reserved.

Oracle is a registered trademark of Oracle Corporation and/or its
affiliates. Other names may be trademarks of their respective
owners.

Type 'help;' or '\h' for help. Type '\c' to clear the current input statement.

#修改xiaoying的客户端ip限制

```
mysql> update mysql.user set host='192.168.1.100' where user='xiaoying';
Query OK, 1 row affected (0.00 sec)
Rows matched: 1  Changed: 1  Warnings: 0
```

#查看用户
```
mysql> SELECT user,host FROM mysql.user;
+---------------+---------------+
| user          | host          |
+---------------+---------------+
| shopdb        | %             |
| xiaogang      | %             |
| shopdb2       | 10.120.71.89  |
| xiaoying      | 192.168.1.100 |
| xiaoming      | 192.168.1.90  |
| mysql.session | localhost     |
| mysql.sys     | localhost     |
| root          | localhost     |
+---------------+---------------+
8 rows in set (0.00 sec)

mysql>
```

9.1.3 删除用户

删除用户使用以下命令。

DELETE FROM mysql.user WHERE user='xiaoying' and host='localhost';

下面将xiaoying用户删除，执行删除命令后，需要执行flush privileges;命令或者重启一下MySQL服务，才能生效。

Microsoft Windows [版本 6.1.7601]
版权所有 (c) 2009 Microsoft Corporation。保留所有权利。

C:\Users\Administrator>mysql –uroot –p123456
mysql: [Warning] Using a password on the command line interface can be insecure.

Welcome to the MySQL monitor. Commands end with ; or \g.
Your MySQL connection id is 19
Server version: 5.7.20 MySQL Community Server (GPL)

Copyright (c) 2000, 2017, Oracle and/or its affiliates. All rights reserved.

Oracle is a registered trademark of Oracle Corporation and/or its
affiliates. Other names may be trademarks of their respective
owners.

Type 'help;' or '\h' for help. Type '\c' to clear the current input statement.

#查看所有用户
```
mysql> SELECT user,host FROM mysql.user;
+---------------+---------------+
| user          | host          |
+---------------+---------------+
| shopdb        | %             |
```

```
| xiaogang        | %            |
| shopdb2         | 10.120.71.89 |
| xiaoying        | 192.168.1.100|
| xiaoming        | 192.168.1.90 |
| mysql.session   | localhost    |
| mysql.sys       | localhost    |
| root            | localhost    |
+-----------------+--------------+
8 rows in set (0.00 sec)
```

#删除用户xiaoying，host为localhost
```
mysql> DELETE FROM mysql.user WHERE user='xiaoying' and host='localhost';
Query OK, 0 rows affected (0.00 sec)
```

#查看用户，xiaoying对应的host是192.168.1.100，并没有删除成功
```
mysql> SELECT user,host FROM mysql.user;
+-----------------+--------------+
| user            | host         |
+-----------------+--------------+
| shopdb          | %            |
| xiaogang        | %            |
| shopdb2         | 10.120.71.89 |
| xiaoying        | 192.168.1.100|
| xiaoming        | 192.168.1.90 |
| mysql.session   | localhost    |
| mysql.sys       | localhost    |
| root            | localhost    |
+-----------------+--------------+
8 rows in set (0.00 sec)
```

#删除用户xiaoying，host为192.168.1.100'，有一条数据受到影响，说明删除有效果
```
mysql> DELETE FROM mysql.user WHERE user='xiaoying' and host='192.168.1.100';
Query OK, 1 row affected (0.00 sec)
```

#查看所有用户，发现xiaoying用户依然存在
```
mysql> SELECT user,host FROM mysql.user;
+-----------------+--------------+
| user            | host         |
+-----------------+--------------+
| shopdb          | %            |
| xiaogang        | %            |
| shopdb2         | 10.120.71.89 |
| xiaoying        | 192.168.1.100|
| xiaoming        | 192.168.1.90 |
| mysql.session   | localhost    |
| mysql.sys       | localhost    |
| root            | localhost    |
+-----------------+--------------+
7 rows in set (0.00 sec)
```

#执行刷新操作
```
mysql> flush privileges;
Query OK, 0 rows affected (0.00 sec)
```

```
#查看所有用户，xiaoying删除成功
mysql> SELECT user,host FROM mysql.user;
+---------------+-----------+
| user          | host      |
+---------------+-----------+
| shopdb        | %         |
| xiaogang      | %         |
| shopdb2       | 10.120.71.89 |
| xiaoming      | 192.168.1.90 |
| mysql.session | localhost |
| mysql.sys     | localhost |
| root          | localhost |
+---------------+-----------+
7 rows in set (0.00 sec)
```

9.2 权限管理

MySQL的用户权限管理主要包括以下几个方面。

（1）设置用户拥有访问数据库、表的权限。

（2）设置用户拥有的操作权限（SELECT、CREATE、UPDATE、DELETE等）。

（3）设置用户使用指定的IP访问权限。

（4）设置用户是否可以给其他用户授权的权限。

精讲视频

权限管理

9.2.1 权限授权语法

授予权限：

GRANT ALL PRIVILEGES ON *.* TO 'xiaogang'@'%' IDENTIFIED BY '123456' WITH GRANT OPTION;

撤销权限：

REVOKE ALL PRIVILEGES ON *.* FROM 'xiaogang'@'%' ;

（1）GRANT：授予权限的关键字。

（2）REVOKE：撤销权限的关键字。

（3）ALL PRIVILEGES：将所有权限授予用户，也可指定具体的权限，如增删改查（SELECT、CREATE、DROP）等权限。

（4）ON：权限对哪些数据库和表生效，格式为数据库名.表名，"*.*"是指对所有数据库和数据表授权。

（5）TO：权限授予用户，格式"用户名"@"登录IP或域名"，%没有限制，在任何主机都可以登录，也可以指定ip或者ip段，"192.168.1.%"，用户只能在192.168.1IP段登录。

（6）FROM：从用户那里回收权限。

（7）IDENTIFIED BY：指定用户的登录密码。

（8）WITH GRANT OPTION：允许用户将自己的权限授予其他用户。

可以使用GRANT给用户添加多个权限，给用户添加了一个SELECT权限后，又给用户添加了一个INSERT权限，该用户就同时拥有了SELECT和INSERT权限。MySQL数据库权限表如表9.1所示。

表9.1　MySQL数据库权限表

Privilege（权限）	Column（字段）	Context（应用范围）	备　　注
ALL [PRIVILEGES]	Synonym for "all privileges"	Server administration	所有权限
ALTER	Alter_priv	Tables	修改表

续表

Privilege（权限）	Column（字段）	Context（应用范围）	备注
ALTER ROUTINE	Alter_routine_priv	Stored routines	更改存储过程、函数
CREATE	Create_priv	Databases, tables, or indexes	创建数据库、表、索引
CREATE ROUTINE	Create_routine_priv	Stored routines	创建存储过程、函数
CREATE TABLESPACE	Create_tablespace_priv	Server administration	创建表空间
CREATE TEMPORARY TABLES	Create_tmp_table_priv	Tables	创建临时表
CREATE USER	Create_user_priv	Server administration	创建用户
CREATE VIEW	Create_view_priv	Views	创建视图
DELETE	Delete_priv	Tables	删除表
DROP	Drop_priv	Databases, tables, or views	删除数据库、表、视图
EVENT	Event_priv	Databases	事件
EXECUTE	Execute_priv	Stored routines	执行
FILE	File_priv	File access on server host	文件
GRANT OPTION	Grant_priv	Databases, tables, or stored routines	授权数据库、表
INDEX	Index_priv	Tables	索引
INSERT	Insert_priv	Tables or columns	插入
LOCK TABLES	Lock_tables_priv	Databases	锁表
PROCESS	Process_priv	Server administration	执行
PROXY	See proxies_priv	table Server administration	代理
REFERENCES	References_priv	Databases or tables	关联
RELOAD	Reload_priv	Server administration	执行flush-hosts, flush-logs, flush-privileges, flush-status, flush-tables, flush-threads, refresh, reload等命令的权限
REPLICATION CLIENT	Repl_client_priv	Server administration	复制权限
REPLICATION SLAVE	Repl_slave_priv	Server administration	复制权限
SELECT	Select_priv	Tables or columns	查询表、字段
SHOW DATABASES	Show_db_priv	Server administration	查看数据库
SHOW VIEW	Show_view_priv	Views	查看视图
SHUTDOWN	Shutdown_priv	Server administration	关闭数据库权限
SUPER	Super_priv	Server administration	执行kill线程权限
TRIGGER	Trigger_priv	Tables	触发器
UPDATE	Update_priv	Tables or columns	更新表、字段
USAGE	Synonym for "no privileges"	Server administration	没有权限

对于数据库表、数据库列以及存储过程应该给予什么权限，官方文档指导说明如表9.2所示。

表9.2 权限配置

权 限 分 布	可能拥有的权限
数据库表权限	Select、Insert、Update、Delete、Create、Drop、Grant、References、Index、Alter
数据库列权限	Select、Insert、Update、References
存储过程权限	Execute、Alter Routine、Grant'

实战演练

下面将创建数据库、表、索引的权限赋予xiaogang这个用户，然后再撤回这个权限，接着授予所有权限。

Microsoft Windows [版本 6.1.7601]
版权所有 (c) 2009 Microsoft Corporation。保留所有权利。

C:\Users\Administrator>mysql -uroot -p123456
mysql: [Warning] Using a password on the command line interface can be insecure.

Welcome to the MySQL monitor. Commands end with ; or \g.
Your MySQL connection id is 5
Server version: 5.7.20 MySQL Community Server (GPL)

Copyright (c) 2000, 2017, Oracle and/or its affiliates. All rights reserved.

Oracle is a registered trademark of Oracle Corporation and/or its
affiliates. Other names may be trademarks of their respective
owners.

Type 'help;' or '\h' for help. Type '\c' to clear the current input statement.

#显示用户xiaogang的权限，USAGE没有任何权限
mysql> SHOW GRANTS FOR xiaogang;
+--+
| Grants for xiaogang@% |
+--+
| GRANT **USAGE** ON *.* TO 'xiaogang'@'%' WITH GRANT OPTION |
+--+
1 row in set (0.00 sec)

#将创建数据库、表、索引的权限授予用户xiaogang
mysql> GRANT CREATE ON *.* TO 'xiaogang'@'%' IDENTIFIED BY '123456' WITH GRANT OPTION;
Query OK, 0 rows affected, 1 warning (0.00 sec)

#显示用户xiaogang的权限，拥有CREATE创建数据库、表、索引的权限
mysql> SHOW GRANTS FOR xiaogang;
+--+
| Grants for xiaogang@% |
+--+
| GRANT CREATE ON *.* TO 'xiaogang'@'%' WITH GRANT OPTION |
+--+
1 row in set (0.00 sec)

#撤销用户xiaogang创建数据库、表、索引的权限
```
mysql> REVOKE CREATE ON *.* FROM 'xiaogang'@'%' ;
Query OK, 0 rows affected (0.00 sec)
```

#权限撤销后，用户xiaogang无任何权限
```
mysql> SHOW GRANTS FOR xiaogang;
+------------------------------------------------------------------+
| Grants for xiaogang@%                                            |
+------------------------------------------------------------------+
| GRANT USAGE ON *.* TO 'xiaogang'@'%' WITH GRANT OPTION |
+------------------------------------------------------------------+
1 row in set (0.00 sec)
```

#授予用户xiaogang所有权限
```
mysql> GRANT ALL PRIVILEGES ON *.* TO 'xiaogang'@'%' IDENTIFIED BY '123456' WITH
 GRANT OPTION;
Query OK, 0 rows affected, 1 warning (0.00 sec)
```

#显示用户xiaogang的权限拥有了所有权限
```
mysql> SHOW GRANTS FOR xiaogang;
+------------------------------------------------------------------+
| Grants for xiaogang@%                                            |
+------------------------------------------------------------------+
| GRANT ALL PRIVILEGES ON *.* TO 'xiaogang'@'%' WITH GRANT OPTION |
+------------------------------------------------------------------+
1 row in set (0.00 sec)
```

#查询用户xiaogang拥有的具体权限
```
mysql> SELECT * FROM mysql.user WHERE user='xiaogang' \G
*************************** 1. row ***************************
                  Host: %
                  User: xiaogang
           Select_priv: Y
           Insert_priv: Y
           Update_priv: Y
           Delete_priv: Y
           Create_priv: Y
             Drop_priv: Y
           Reload_priv: Y
         Shutdown_priv: Y
          Process_priv: Y
             File_priv: Y
            Grant_priv: Y
       References_priv: Y
            Index_priv: Y
            Alter_priv: Y
          Show_db_priv: Y
            Super_priv: Y
 Create_tmp_table_priv: Y
      Lock_tables_priv: Y
          Execute_priv: Y
       Repl_slave_priv: Y
```

```
             Repl_client_priv: Y
             Create_view_priv: Y
               Show_view_priv: Y
          Create_routine_priv: Y
           Alter_routine_priv: Y
             Create_user_priv: Y
                   Event_priv: Y
                 Trigger_priv: Y
       Create_tablespace_priv: Y
                     ssl_type:
                   ssl_cipher:
                  x509_issuer:
                 x509_subject:
                max_questions: 0
                 max_updates: 0
              max_connections: 0
         max_user_connections: 0
                       plugin: mysql_native_password
        authentication_string: *6BB4837EB74329105EE4568DDA7DC67ED2CA2AD9
             password_expired: N
        password_last_changed: 2018-02-02 14:37:49
            password_lifetime: NULL
               account_locked: N
1 row in set (0.00 sec)

mysql>
```

9.2.2 账户权限体系

账户权限体系分为服务级用户权限、数据库级用户权限、表级用户权限、字段级用户权限，不同的用户级别拥有不同的权限。

1. 服务级用户权限

服务级用户权限拥有对所有数据库进行操作的权限，如同root用户一样，可以删除所有数据库以及表，权限存储在mysql.user表中。GRANT ALL ON *.*和REVOKE ALL ON *.*只授予和撤销全局权限，*.*表示所有数据库、所有数据表。

#授予用户xiaogang为服务级用户权限，具有grant 权限，可以创建其他用户的权限
GRANT ALL PRIVILEGES ON *.* TO 'xiaogang'@'%' IDENTIFIED BY '123456' WITH GRANT OPTION;

2. 数据库级用户权限

在具有数据库级用户权限的数据库中可以进行增删改查等操作（依据分配的权限），权限存储在mysql.db和mysql.host表中，GRANT ALL ON db_name.*和REVOKE ALL ON db_name.*只授予和撤销数据库级权限，db_name.*表示指定的数据库。

#授予用户xiaogang数据库级用户权限，可以操作staff数据库的所有权限
GRANT ALL PRIVILEGES ON staff.* TO 'xiaogang'@'%' IDENTIFIED BY '123456' WITH GRANT OPTION;

3. 表级用户权限

表级用户权限即对给定表中的所有列进行操作的权限（依据分配的权限），权限存储在mysql.tables_priv表中，GRANT ALL ON db_name.tbl_name和REVOKE ALL ON db_name.tbl_name只授予和撤销表权限，db_name.tbl_name表示指定的数据库和指定的表。

#授予用户xiaogang表级用户权限，可以对staff数据库中的employee表进行操作的所有权限
GRANT ALL PRIVILEGES ON staff.employee TO 'xiaogang'@'%' IDENTIFIED BY '123456' WITH GRANT OPTION;

4. 字段级用户权限

字段级用户权限是对给定表中的单一列操作的权限（依据分配的权限），权限存储在mysql.columns_priv表中，当使用REVOKE时，必须指定与被授权列相同的列。字段级用户权限使用频率低，每次访问时都需要权限校验，效率低下。

9.2.3 权限授予原则

在进行信息安全检查或者信息系统安全等级保护测评的过程中，需要对数据库进行安全扫描，如果权限授予有问题，就会产生漏洞。下面列举一些常见的数据库漏洞。

（1）super 权限授予了除 root 外的用户。

风险等级：中。

产生原因：super 权限授予了除 root 外的用户。

漏洞描述：该策略检测super的权限是否授予了除root外的用户。拥有super权限的用户可以终止其他用户的mySQL进程。在一个安全的环境中，super权限只授予root用户。

修复建议：回收授予的除 root 外的用户权限。命令为revoke super on *.* from '{user}'@'{host}';其中参数 user 为需要修改的用户名，host 为用户所在主机名。

（2）create user 权限授予了除 root 外的用户。

风险等级：中。

产生原因：create user 的权限授予了除 root 外的用户。

漏洞描述：该策略检测create user的权限是否授予了除root外的用户。拥有创建用户权限的用户可以在数据库服务器中创建用户，并且可以把对数据库操作的权限赋予新创建的用户，如select、insert、update、delete，也就是说得到了创建用户权限的用户可以间接控制数据库服务器。在一个安全的环境中，create user的权限只授予root用户。

修复建议：回收授予的除 root 外的用户权限。命令为revoke create user on *.* from '{user}'@'{host}';其中参数 user 为需要修改的用户名，host 为用户所在主机名。

（3）存在 file 全局权限用户。

风险等级：中。

产生原因：存在 file全局权限的普通用户。

漏洞描述：该策略检测是否存在file全局权限的普通用户。如用户被授予file全局权限，则可以通过LOAD DATA INFILE和SELECT...INTO OUTFILE语句读写服务器上的任何文件。从数据库安全角度考虑，只有管理员用户才被授予file全局权限。

修复建议：通过命令update mysql.user set file_priv='N' where user='{user}' and host='{host}'撤销用户的 file全局权限；其中参数 user 为需要修改的用户名，host 为用户所在主机名。

（4）存在 drop 全局权限用户。

风险等级：中。

产生原因：存在 drop 全局权限的普通用户。

漏洞描述：该策略检测是否存在drop全局权限的普通用户。如用户被授予drop全局权限，则可能会有重要数据库或表被删除的危险。从数据库安全角度考虑，只有管理员用户才被授予drop全局权限。

修复建议：通过命令drop user '{user}'@'{host}'删除拥有 drop 全局权限的普通用户；或通过命令update mysql.user set drop_priv='N' where user='{user}' and host='{host}'撤销普通用户的 drop全局权限；其中参数 user 为需要修改的用户名，host 为用户所在主机名。

（5）存在 grant 数据库级权限用户。

风险等级：中。

产生原因：存在 grant 数据库级权限的普通用户。

漏洞描述：该策略检测是否存在grant数据库级权限的普通用户。grant数据库级权限允许用户在全局范围内授予其他用户权限。在某些版本中，具有grant权限的用户可以重置其他任何用户的密码，这可能会导致一个用户通过修改其他所有用户密码而控制整个数据。

修复建议：通过命令update mysql.db set grant_priv='N' where user='{user}' and host='{host}'撤销普通用户对 grant 的数据库级权限；其中参数 user 为需要修改的用户名，host 为用户所在主机名。

（6）存在 drop 数据库级权限用户。

风险等级：中。

产生原因：存在 drop 数据库级权限的普通用户。

漏洞描述：该策略检测是否存在drop数据库级权限的普通用户。如用户被授予drop数据库级权限，则可能会有重要数据库或表被删除的危险。从数据库安全角度考虑，只有管理员用户才被授予drop数据库级权限。

修复建议：通过命令update mysql.db set drop_priv='N' where user='{user}' and host='{host}'撤销普通用户对 drop 的数据库级权限；其中参数 user 为需要修改的用户名，host 为用户所在主机名。

（7）存在 process 全局权限用户。

风险等级：中。

产生原因：存在 process 权限的用户。

漏洞描述：该策略检测是否存在process权限的用户。process是管理权限，process权限能被用来查看当前执行查询的明文文本，包括设定或改变密码的查询。从数据库安全角度考虑，只有管理员用户才可以被授予process权限。

修复建议：通过命令update mysql.user set process_priv='N' where user='{user}' and host='{host}'撤销用户的 process 权限；其中参数 user 为需要修改的用户名，host 为用户所在主机名。

（8）从任意主机登录的用户。

风险等级：中。

产生原因：存在从任意主机都能登录数据库的用户。

漏洞描述：该策略检测是否存在从任意主机都能登录数据库的用户。当用户请求连接时，服务器会用user表中的host、user、password 3个字段进行身份验证，其中host为用户所在主机，如果user表中某一用户的host字段值为空或者为通配符"%"，则意味着该用户可以从任意主机登录数据库，从安全角度来考虑这是不允许的。

修复建议：通过命令update mysql.user set host='{newhost}' where user='{username}'and host in('','%'); flush privileges修改 user 的 host 字段；其中参数 newhost 表示指定的主机 ip 或者主机名，username表示需要修改的用户名。

（9）存在拥有 replication slave 权限的普通用户。

风险等级：低。

产生原因：存在拥有 replication slave 权限的普通用户。

漏洞描述：该策略检测是否存在拥有replication slave权限的普通用户。如用户被授予replication slave权限，则可以查看从服务器，从主服务器读取二进制日志。因此，建议该权限只授权给真正需要该权限的用户。

修复建议：通过命令：drop user '{user}'@'{host}'删除拥有 replication slave 权限的普通用户；或通过命令update mysql.user set repl_slave_priv='N' where user='{user}' and host='{host}'撤销普通用户的replication slave 权限；其中参数 user 为需要修改的用户名，host 为用户所在主机名。

（10）存在拥有 show database 权限的普通用户。

风险等级：低。

产生原因：存在拥有 show database 权限的普通用户。

漏洞描述：该策略检测是否存在拥有show database权限的普通用户。如用户被授予show database权限，则可以查看数据库服务器中所有的数据库，得到数据库服务器的敏感信息。因此，建议该权限只授权给真正需要该权限的用户。

修复建议：通过命令drop user '{user}'@'{host}'删除拥有 show database 权限的普通用户，或通过命令update mysql.user set show_db_priv='N' where user='{user}' and host='{host}'撤销普通用户的 show database 权限；其中参数 user 为需要修改的用户名，host 为用户所在主机名。

从以上10个MySQL数据库漏洞，可以总结出以下权限授予原则。

（1）不要授予普通用户super 权限、create user 权限、drop 全局权限、grant 数据库级权限等这些管理员所拥有的权限。

（2）授予普通用户最小权限，权限够用就行，如用户需要增删改查的权限，就不要授予其删除用户的权限。

（3）数据库用户可以限制登录的主机，可以指定IP或者IP段。

（4）用户的密码要足够复杂，可以用字母、数字、特殊字符、大小写组成复杂密码。

（5）清理不用的数据库和用户，减少数据库使用空间以及不用的用户。

9.3 表空间管理（InnoDB）

MySQL数据库采用InnoDB存储引擎，就会有两种表空间可以使用：一种是共享表空间；另外一种是独占表空间。

（1）共享表空间。共享表空间是指一个数据库的所有表数据以及索引文件全部放在一个文件中。共享表空间的默认文件路径在data目录下，默认的文件名为ibdata1，初始化为10M，如图9.1所示。

共享表空间可以将表空间分成多个文件存放到各个磁盘上，数据和文件放在一起方便管理，同时数据和文件都放到一个文件里会产生一个很大的文件，对一个表做了大量删除操作后其表空间中将会有大量的空隙，这样会浪费空间。

（2）独占表空间。独占表空间是将每个表都以独立的文件方式来进行存储，每个表都有一个frm、ibd表文件。frm文件保存了每个表的元数据，包括表结构的定义等；ibd文件存放该表的数据和索引文件，如图9.2所示。

精讲视频

表空间管理（InnoDB）

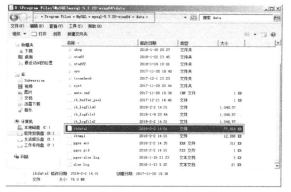

图9.1　共享表空间文件

图9.2　独占表空间文件

独占表空间使每一个数据库的每一个表都有自己独立的表空间，每个表的数据和索引都会存放在自己的表空间中，可以实现单表在不同的数据库中移动。与共享表空间相比，使用独占表空间的效率以及性能会更高一点。

（3）开启表空间。默认情况下，数据库使用的表空间为共享表空间，但是可以通过参数innodb_file_per_table 来将其设置为独占表空间，如果值为OFF说明所使用的是独占表空间。

```
mysql> show variables like "innodb_file_per_table";
+-----------------------+-------+
| Variable_name         | Value |
+-----------------------+-------+
| innodb_file_per_table | ON    |
+-----------------------+-------+
1 row in set, 1 warning (0.00 sec)
```

在配置文件my.ini或者my.cnf里配置innodb_file_per_table=1 是否使用共享以及独占表空间，1为使用独占表空间，0 为使用共享表空间。

（4）查看数据库共享表空间。使用show variables like 'innodb_data%';查看共享表空间的大小，ibdata1的默认大小为10M，超过10M可以自动扩展。innodb_data_file_path可以进行多路径存储设置，innodb_data_file_path = /data1/db1/ibdata1:100M:autoextend; /data2/db2/ibdata2:100M:autoextend 放在不同的磁盘中，可以平均磁盘负载，提高数据库性能。

```
mysql> show variables like 'innodb_data%';
+-----------------------+--------------------+
| Variable_name         | Value              |
+-----------------------+--------------------+
| innodb_data_file_path | ibdata1:12M:autoextend |
| innodb_data_home_dir  |                    |
+-----------------------+--------------------+
2 rows in set, 1 warning (0.00 sec)
```

（5）查看所有数据库占用空间大小。可以查看各个数据库所占空间的大小以及索引文件的大小。

```
mysql> select TABLE_SCHEMA, concat(truncate(sum(data_length)/1024/1024,2),' MB')
    as data_size,
    -> concat(truncate(sum(index_length)/1024/1024,2),'MB') as index_size
    -> from information_schema.tables
    -> group by TABLE_SCHEMA
    -> order by data_length desc;
+--------------------+------------+------------+
| TABLE_SCHEMA       | data_size  | index_size |
+--------------------+------------+------------+
| staff2             | 834.29 MB  | 0.00MB     |
| staff              | 300.89 MB  | 0.00MB     |
| bsweb              | 2572.80 MB | 37.69MB    |
| tzoawhweb          | 0.45 MB    | 0.09MB     |
| shop               | 0.28 MB    | 0.04MB     |
| information_schema | 0.15 MB    | 0.00MB     |
| performance_schema | 0.00 MB    | 0.00MB     |
| mysql              | 2.27 MB    | 0.21MB     |
| sys                | 0.01 MB    | 0.00MB     |
+--------------------+------------+------------+
9 rows in set (0.25 sec)
```

通过统计数据库的占用大小，可以看到这个数据库的空间使用以及索引文件的大小，如员工管理系统数据库staff的空间占用是300.89 MB，索引文件是0.00 MB。

（6）查看指定数据库占用空间大小。可以查看指定数据库的各个表所占空间大小以及表索引文件的大小。

```
mysql> select TABLE_NAME, concat(truncate(data_length/1024/1024,2),' MB') as dat
a_size,
    -> concat(truncate(index_length/1024/1024,2),' MB') as index_size
    -> from information_schema.tables where TABLE_SCHEMA = 'staff'
    -> group by TABLE_NAME
    -> order by data_length desc;
+------------+-----------+------------+
| TABLE_NAME | data_size | index_size |
+------------+-----------+------------+
| employee   | 118.64 MB | 0.00 MB    |
| ask_leave  | 111.62 MB | 0.00 MB    |
| payroll    | 70.59 MB  | 0.00 MB    |
| dept       | 0.01 MB   | 0.00 MB    |
+------------+-----------+------------+
5 rows in set (0.00 sec)
```

通过查看指定数据库空间的使用，可以看到各个表的空间使用情况和索引情况，如员工管理系统数据库staff的员工表employee的空间占用是111.64 MB，索引文件大小为0.00 MB，而部门表dept的空间占用是0.01 MB，索引文件大小为0.00 MB。

9.4 备份与还原

精讲视频

备份与还原

MySQL数据库备份与还原也是在项目上线后必须要做的一件事。数据库运行过程中，很有可能遇到停电、磁盘损坏、数据库服务器宕机、自然灾害、黑客攻击等问题，这些都有可能使数据丢失，那么对数据库的备份就显得格外重要了，定期备份可以防止数据丢失。

9.4.1 备份数据

MySQL数据库备份可以分为热备份、温备份、冷备份。热备份是指当数据库进行备份时，数据库的读写操作不受影响；温备份是指当数据库进行备份时，数据库的读操作可以进行，但是不能进行写操作；冷备份是指当数据库进行备份时，数据库不可以进行读写操作。

备份的方式可以分为物理备份和逻辑备份。

物理备份是直接拷贝数据库文件，包括数据、二进制日志、InnoDB事务日志、代码（存储过程、存储函数、触发器、事件调度器）、服务器配置文件。如果数据库是独占表空间，可以直接拷贝要备份的数据库文件，如备份员工管理系统staff，可以直接拷贝staff整个文件夹如图9.3所示；如果是共享表空间，需要拷贝数据文件ibdata1、日志文件ib_logfile，如图9.4所示。

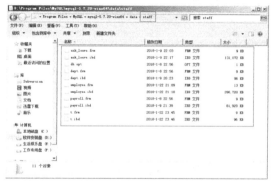

图9.3　独占表空间数据库物理备份

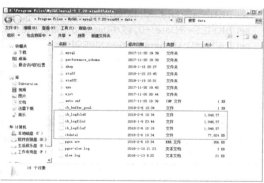

图9.4　共享表空间数据库物理备份

逻辑备份可以采用MySQL数据库自带的备份工具mysqldump，它可以备份成指定的文本文件或者可执行SQL脚本文件；或者用第三方的xtrabackup备份，它是一款非常强大的InnoDB/XtraDB热备工具，支持完全备份、增量备份。

下面使用MySQL数据库自带的备份工具mysqldump进行备份，将员工管理系统数据库staff备份出来。

```
#备份员工管理系统数据库staff
C:\Users\Administrator>mysqldump -uroot --password=123456 staff > E:\mysql_bak\staff_20180206.sql

#备份192.168.1.90服务器上的员工管理系统数据库staff
C:\Users\Administrator>mysqldump --opt -uroot --password=123456 -h192.168.1.90 staff > E:\mysql_bak\staff_20180206.sql

#备份员工管理系统数据库staff中的员工表employee
C:\Users\Administrator>mysqldump -uroot --password=123456 staff employee> E:\mysql_bak\staff_20180206.sql
```

（1）mysqldump：备份关键字。

（2）--opt：可选项，如果加上--opt参数，建表语句包含drop table if exists tableName，insert之前包含一个锁表语句lock tables tableName write，insert之后包含unlock tables。

（3）-u：用户名。

（4）--password：用户密码。

（5）-h：可以指定远程数据库的IP地址。

（6）staff：要备份的数据库名称。

（7）E:\mysql_bak\staff_20180206.sql：备份的文件路径。

（8）--default-character-set：设置字符集。

（9）--single-transaction：将导出设置成事件。

（10）--no-data：导出的SQL脚本中，将只包含创建表的create 语句。

（11）--add-drop-table：导出的脚本中，包含 drop table if exists。

（12）--routines：导出存储过程及函数。

（13）--events：导出事件。

（14）--triggers：导出触发器。

mysqldump通用的备份脚本如下所示。

```
@echo off
set BAT_HOME=E:\mysql_bak
set DaysAgo=30
forfiles /p %BAT_HOME% /s /m *.* /d -%DaysAgo% /c "cmd /c del @path"
set BKDIR=%Date:~0,4%%Date:~5,2%
set BKFILE=%Date:~0,4%%Date:~5,2%%Date:~8,2%
set PATH=D:\Program Files\MySQL\mysql-5.7.20-winx64\bin
mkdir %BAT_HOME%\%BKDIR%
mysqldump --opt -u root --password=123456 -h192.168.1.90 staff > %BAT_HOME%\%BKDIR%\staff_%BKFILE%.sql
echo "数据库备份完成！
```

员工管理系统数据库staff的备份结果如图9.5所示。

9.4.2 还原数据

通过物理备份方式进行备份的数据库，可以将备份出来的文件拷贝到数据库安装的data目录下，如D:\Program Files\MySQL\mysql-5.7.20-winx64\data，来还原数据库。

第9章 MySQL管理

图9.5 数据库备份结果

通过使用MySQL数据库自带的mysqldump备份工具进行备份的数据库，有多种方式进行还原，可以使用MySQL客户端还原备份出来的sql文件，也可以使用命令来进行还原，如下所示。

```
mysql> USE staff
mysql> source E:/mysql_bak/201802/staff_20180201.sql;
```

9.5 主从同步配置

主从同步配置

数据库在使用过程中，为了保证数据的安全性，往往会进行备份操作，这时就可以使用主从数据库同步配置，将主数据库中的数据同步到从数据库里，以达到备份的目的。当数据库读写操作频繁，用户访问量较大时，让主数据库进行写入操作，从数据库进行读取操作，以实现数据库读写分离操作，可以缓解数据库的压力。主从同步配置需要两个MySQL数据库，它们最好分布在两台服务器上。下面学习在Windows操作系统上对数据库进行主从同步配置。

9.5.1 主数据库配置

主数据库配置需要在my.ini配置文件中添加参数、分配用于同步的用户、查找同步的二进制日志文件和位置。

（1）打开MySQL数据库的my.ini配置文件，添加数据库唯一标识server-id（主数据库和从数据库需要不一致），开启log-bin日志以及需要同步的数据库binlog-do-db，binlog-ignore-db用来配置不需要同步的数据库。

```
server-id=1                    #主数据库和从数据库需要不一致
log-bin=mysql-bin              #开启log-bin二进制日志文件
binlog-do-db=db_test           #需要同步的数据库
binlog-ignore-db=staff         #不需要同步的数据库
```

（2）查看log-bin日志文件是否开启成功，只有log-bin等于ON，文件开启成功，才可以进行同步操作。

```
mysql> show variables like 'log_bin%';
+---------------------------------+-----------------------------------------------------------------+
| Variable_name                   | Value                                                           |
+---------------------------------+-----------------------------------------------------------------+
| log_bin                         | ON                                                              |
| log_bin_basename                | D:\Program Files\MySQL\mysql-5.7.20-winx64\data\mysql-bin       |
| log_bin_index                   | D:\Program Files\MySQL\mysql-5.7.20-winx64\data\mysql-bin.index |
| log_bin_trust_function_creators | OFF                                                             |
| log_bin_use_v1_row_events       | OFF                                                             |
+---------------------------------+-----------------------------------------------------------------+
```

（3）分配用于同步的账号backup，密码是123456。

```
GRANT REPLICATION SLAVE ON *.* to 'backup'@'%' IDENTIFIED BY '123456';
```

（4）查看同步的二进制日志文件名称mysql-bin.000002和位置437，用于在进行从数据库同步配置时使用。

```
mysql> show master status;
+------------------+----------+--------------+------------------+-------------------+
| File             | Position | Binlog_Do_DB | Binlog_Ignore_DB | Executed_Gtid_Set |
+------------------+----------+--------------+------------------+-------------------+
| mysql-bin.000002 |      437 | db_test      |                  |                   |
+------------------+----------+--------------+------------------+-------------------+
```

9.5.2 从数据库配置

从数据库配置需要在my.ini配置文件中添加参数、配置同步的主数据库、开启同步的从数据库。

（1）打开MySQL数据库的my.ini配置文件，添加数据库唯一标识server-id（主数据库和从数据库需要不一致），开启log-bin日志以及需要同步的数据库replicate-do-db，并把从数据库设置为只读数据库。

```
server-id=2                    #主数据库和从数据库需要不一致
log-bin=mysql-bin              #开启log-bin二进制日志文件
replicate-do-db=db_test        #需要同步的数据库
read_only                      #为保证数据库的数据一致性，从数据库只允许读取操作，不允许写操作
```

（2）配置同步的数据库，MASTER_HOST是主数据ip，MASTER_PORT是主数据库端口号，MASTER_USER是主数据库分配的同步账号，MASTER_LOG_FILE是主数据库同步的二进制日志文件，MASTER_LOG_POS是同步文件的位置。

CHANAGE MASTER TO MASTER_HOST='192.168.1.90',MASTER_PORT=3306,MASTER_USER='backup',MASTER_PASSWORD='123456',MASTER_LOG_FILE='mysql-bin.000002',MASTER_LOG_POS=437;

（3）开启从数据库同步START SLAVE。

mysql> START SLAVE;

（4）查看从数据库同步状态SHOW SLAVE STATUS，查看同步的二进制日志文件名称mysql-bin.000002和位置437，Slave_IO_Running等于Yes和Slave_SQL_Running等于Yes说明同步配置成功，可以用于主从数据库同步。

```
mysql> SHOW SLAVE STATUS\G
*************************** 1. row ***************************
            Slave_IO_State: Waiting for master to send event
               Master_Host: 192.168.1.90          #主数据库的ip
               Master_User: backup                #主数据库分配的同步账号
               Master_Port: 3306                  #主数据库端口号
             Connect_Retry: 60
           Master_Log_File: mysql-bin.000002      #主数据库同步的二进制日志文件
       Read_Master_Log_Pos: 437                   #主数据库同步的二进制日志文件的位置
            Relay_Log_File: WINDOWS-MKCKMF8-relay-bin.000002
             Relay_Log_Pos: 320
     Relay_Master_Log_File: mysql-bin.000002
          Slave_IO_Running: Yes
         Slave_SQL_Running: Yes
           Replicate_Do_DB:db_test    #要同步的数据库
       Replicate_Ignore_DB:
        Replicate_Do_Table:
    Replicate_Ignore_Table:
   Replicate_Wild_Do_Table:
Replicate_Wild_Ignore_Table:
                Last_Errno: 0
```

```
                  Last_Error: 
               Skip_Counter: 0
        Exec_Master_Log_Pos: 643
            Relay_Log_Space: 537
            Until_Condition: None
             Until_Log_File: 
              Until_Log_Pos: 0
         Master_SSL_Allowed: No
         Master_SSL_CA_File: 
         Master_SSL_CA_Path: 
            Master_SSL_Cert: 
          Master_SSL_Cipher: 
             Master_SSL_Key: 
      Seconds_Behind_Master: 0
Master_SSL_Verify_Server_Cert: No
              Last_IO_Errno: 0
              Last_IO_Error: 
             Last_SQL_Errno: 0
             Last_SQL_Error: 
 Replicate_Ignore_Server_Ids: 
           Master_Server_Id: 1
                Master_UUID: 762a6f83-cde7-11e7-89fa-8056f2d6b9b6
           Master_Info_File: D:\Program Files\MySQL\mysql-5.7.21-winx64\data\master.info
                  SQL_Delay: 0
        SQL_Remaining_Delay: NULL
    Slave_SQL_Running_State: Slave has read all relay log; waiting for more updates
         Master_Retry_Count: 86400
                Master_Bind: 
    Last_IO_Error_Timestamp: 
   Last_SQL_Error_Timestamp: 
            Master_SSL_Crl: 
         Master_SSL_Crlpath: 
         Retrieved_Gtid_Set: 
          Executed_Gtid_Set: 
              Auto_Position: 0
        Replicate_Rewrite_DB: 
               Channel_Name: 
        Master_TLS_Version: 
```

（5）在主数据库上创建要同步的数据库db_test，可以在从数据库中使用SHOW DATABASES命令查看到其已经同步到了从数据库里，进一步说明主从数据库同步配置成功。

```
mysql> SHOW DATABASES;
+--------------------+
| Database           |
+--------------------+
| information_schema |
| mysql              |
| performance_schema |
| db_test            |
| sys                |
+--------------------+
```

9.6 小结

本章讲解了MySQL数据库的管理功能，数据库的用户管理，如何创建用户、修改用户以及删除用户；数据库的权限管理，权限授权语法、账户权限体系以及权限授予原则；表空间管理，共享表空间和独占表空间的使用；备份与还原管理，如何备份数据库以及还原数据库；主从同步配置，用来进行数据库备份操作或者数据库读写分离操作。

拓展视频： MySQL高性能优化。

为什么查询慢　　数据准备　　慢查询日志分析　　索引优化

特定类型优化　　数据库表优化　　数据库配置优化　　服务器硬件优化